Journeys in the American Pacific: A Travel Memoir

JOURNEYS IN THE AMERICAN PACIFIC: A TRAVEL MEMOIR

PETER W NOONAN

MAGISTRALIS
OTTAWA, ONTARIO

Journeys in the American Pacific: A Travel Memoir Copyright © 2023 by Peter W Noonan. All Rights Reserved.

Published by Magistralis, Ottawa, Canada

ISBN (hardcover) 978-1-7780030-1-1

Cataloging in Publication Data for this book can be obtained from:

Library and Archives Canada, 395 Wellington Street, Ottawa, ON K1A 0N4 Canada

Contents

Dedication vii
Image Credits ix
Author's Note xi
Preface xiii

Part I.
THE HAWAIIAN ISLANDS

1. Introduction to the Hawaiian Islands 17
2. Oahu - The Island of History 23
3. Pearl Harbour - The Birth of a World Power 39
4. Kauai - Hawaii's Garden Paradise 45

Part II.
THE ISLANDS OF MICRONESIA

5. Introduction to Micronesia 51
6. Yap – The Island of Tradition 61
7. Truk Lagoon - The Sunken Museum 99
8. Pohnpei - The Capital of an Emerging Island Nation 103
9. The Mysterious City of Nan Madol 123
10. Palau - The Last Trust Territory 137
11. Saipan - The Island of Bitter Memory 155
12. Guam - America in the Western Pacific 167

Part III.
RETURN TO THE LAND OF THE KAMEHAMEHA'S

13. Oahu and Hawaii Island 187
14. A Final Reflection 231

DEDICATION

To the people of these islands

IMAGE CREDITS

All photographs are credited to the author unless otherwise indicated

AUTHOR'S NOTE

Other than historical figures, the names of the people that are mentioned in this book have been changed with the exception of Francis Torobiong, the divemaster that I met in Palau.

PREFACE

Ah, the Pacific Ocean! That vast expanse of blue water that occupies so much of the surface of our planet, speckled here and there by the emerald green spires of tropical islands, and the powder blue waters of calm lagoons encircled by white sand beaches. For centuries they have been everyone's image of a paradise on Earth. As a boy growing up in Canada in the Sixties my imagination was captured by tales of those islands as I watched the television actor Gardner McKay pilot his schooner, the *Tiki*, through the South Seas on the syndicated television serial *Adventures in Paradise*. I devoured books on the South Seas and in doing so I sailed with Fletcher Christian and the *Bounty* mutineers in Nordoff and Hall's *Mutiny on the Bounty*, a fictionalized account of Captain Bligh's tyranny amidst a voyage to fabled Tahiti. Still later, I found myself on the shores of a Pacific idyll with Herman Melville in *Typee*, his novelistic recounting of his sojourn with the fierce warriors of the Typee Valley on Nuka Hiva, in the cannibalistic Marquesas Islands. James Michener's *Tales of the South Pacific* described how the great Pacific war of the twentieth century affected many of the oceanic islands, and altered the lives of the men and women who were there. In its sequel, *Return to Paradise*, he captured a snapshot of the final stages of colonialism in the Pacific. Michener's great masterpiece, however, was *Hawaii*, his tale of the fabled island kingdom and its slow fall into the grasp of the United States, and the subsequent fate of the Hawaiian people. Those were the stories of kings and common folk, European and American adventurers, settlers from the Orient, and the beautiful South Seas women who came to love them. I could imagine myself walking along the shores of one of these paradisiacal islands in an idyllic escape as I pored over National Geographic Society maps of the Pacific Ocean, searching for many of the tiny islands described in the books that I had read.

Green isles and blue waters of the Pacific Ocean

As I came to adulthood the romanticism of the Pacific was pushed into the background by the more prosaic duties and obligations of modern life. But I never entirely lost the lure of the Pacific islands and when I was able to do so I chose to visit some of those islands, hoping that they might yet retain some of the authenticity portrayed in the South Seas literature that I read in my youth. What follows is an account of my personal journeys to some of those fabulous islands of the Pacific, rich in culture and history, and forever part of our imagination of paradise.

The islands that I visited in this travel memoir are linked by a common theme. They have all been profoundly affected in different ways by their relationship with the United States, which over the course of its long history has become the great hegemonic power in the North Pacific Ocean in which these islands are found. For some of these islands, their dance with the United States proved fatal, or nearly so, while for others long periods of beneficial neglect under the watchful eye of the United States have ensured the survival and continuity of much of their ancient cultures. Nevertheless, the title of the book is not a reflection on the present independence and sovereignty of any of the nations that I visited in Micronesia. When I visited Micronesia in 1989, Palau had not yet emerged from the US Trust Territory of the Pacific Islands and while the Federated States of Micronesia and the Marshall Islands had been granted independence by the United States it was not until 1990 that the United Nations formally recognized the termination of the trust in relation to those islands.

Across the Pacific, I have encountered people in Micronesian islands who live almost as their forefathers did a century or more ago, and people in Hawaii who live at the leading edges of twenty-first-century life. All of them remain linked by the Pacific Ocean, the great mother of their islands, peaceful, tempestuous, sometimes explosive, but always the ever-present great blue sea that gives life to their emerald islands (and sometimes takes it away). To the people of those islands whose lives have touched mine, ever so briefly, this account is respectfully dedicated.

PART I
THE HAWAIIAN ISLANDS

INTRODUCTION TO THE HAWAIIAN ISLANDS

The Hawaiian Islands are uniquely located on the surface of the Earth. As a landmass, the Hawaiian Islands are farther away from any other landmass on the planet. The Hawaiian Islands are also the longest archipelago in the world, stretching across more than 2400 kilometres of the North Pacific Ocean. Born of the fires of magma beneath the floor of the Pacific Ocean the current islands rose in succession through enormous volcanic eruptions that raised mountains high above the ocean. Subsequently, the islands raised by volcanic eruptions lost their volcanic engines through the continual movement of the Earth's tectonic plates, and as those plates shifted the Hawaiian Islands north-westwards they slowly shrank, until they became mere atolls, or reefs, within the great South Seas. Today, the large inhabited islands lie at the southernmost end of the island chain while the north-western islands are now small uninhabited specks in the ocean, a remnant of the high islands that they once were.

Map of the Principal Islands of Hawaii (Wikimedia Commons, Courtesy of the University of Texas Libraries, The University of Texas at Austin, Perry-Castañeda Library Map Collection)

The relative isolation of the Hawaiian archipelago also meant that it was among the last places in the world to receive the blessing, or curse, of human habitation. Archaeologists posit that Hawaii was initially settled by mariners from the Marquesas Islands far to the south, in what is now the territory of French Polynesia. Sometime between 300 and 400 AD adventurous mariners pointed their double-hulled ocean-going canoes north and set out on a perilous and monumental journey that took them to the Hawaiian archipelago where they first landed on the youngest, and largest, of the Hawaiian chain, what we call today Hawaii Island or, more colloquially, the Big Island. There the Marquesans established a vibrant Polynesian culture that subsequently spread to the other major islands of the Hawaiian chain in a migration that was refreshed by further waves of immigration from other southern Polynesian islands, principally Tahiti. In splendid isolation, the population of the islands grew until the population reached perhaps 680,000 people by the late 1770s.

Then, in 1778, an epochal event occurred that forever changed Hawaii. The British navigator and explorer Captain James Cook arrived with his vessels of exploration, *HMS Resolution*, and *HMS Discovery*, and thereafter Hawaii became known to the wider world. Although there is historical cartographic evidence that Spanish treasure galleons crossing the Pacific Ocean from South America to the Philippines, and laden with the silver treasure that was mined in the Andes

Mountains, discovered Hawaii first there is no substantial record of contact between the Spanish and the Hawaiians. Thus, it was the British, and more particularly Captain James Cook, who put Hawaii on the map of the world. Cook's ships visited Hawaii, which he called the Sandwich Islands, three times but unfortunately, those visits began the process of the decimation of the Hawaiian people, and the consequent undermining of their culture through the introduction of a whole range of infectious bacteriological and viral diseases, principally respiratory and venereal diseases. On top of the introduction of diseases for which the population had no natural immunity, the population of the islands suffered a form of culture shock as the Hawaiians came face to face with a civilisation that was technologically far superior in comparison to the civilisation of Polynesia and one that had very different conceptions of morality and behaviour. In fact, it was a cultural misunderstanding about property rights over a boat that ultimately led to the death of Captain Cook on Hawaii Island. But the long-term damage to the Hawaiian nation from first contact with the west was much greater than the mere loss of a boat.

At the time of first contact between the British and the Hawaiians, a native man of unusual ability rose to the challenge of western contact by fundamentally altering the course of the Hawaiian Islands through military conquest and political organization. That man was Kamehameha, a towering figure in Hawaiian history both physically and historically. Seizing a destiny that had been foretold he conquered his native Hawaii Island and then obtained western weapons that allowed him to invade and militarily subjugate all of the Hawaiian Islands except for Kauai, and its small neighbour Niihau, both of which lay far to the north-west of Hawaii Island. In 1795, Kamehameha established a monarchical government over the islands that he controlled and the Kingdom of Hawaii came into existence. Kamehameha's kingdom allowed the western political states of the era to treat the Hawaiian government as an organized polity. Fifteen years later, in 1810, King Kamehameha obtained the peaceful allegiance of the two northernmost inhabited islands of the chain, Kauai and Niihau, and a unified kingdom of all of the major Hawaiian Islands took on the form that it would retain until the end of the nineteenth century.

But the tide of westernization of the islands could not be withstood, although King Kamehameha I strove mightily to consolidate his kingdom and to integrate it into the wider world. Upon his death, however, his wives and son threw over the traditional Hawaiian cosmogony and religion of the islands and the coincidental arrival of a group of Calvinist missionaries from the United States thrust Hawaii down the path of Christianity and westernization. A long-term decline in the native population as a result of the introduction of successive waves of infectious diseases, for which the people had no natural resistance, a resulting loss of native fecundity, and the expanding tentacles of western commerce, all compromised the stability and independence of the Kingdom. In 1887, a revolt by western plantation owners and commercial investors considerably reduced the power of the Hawaiian monarch and spurred an American annexation movement among a group of Hawaiian-born and Hawaiian-naturalized citizens, who primarily traced their family histories back to the United States, and who considered themselves to be American. In 1893, the annexationists struck, taking advantage of an ill-advised attempt by the then-current monarch, Queen Liliuokalani, to restore some of the powers and prestige of the monarchy that had been taken away by the revolt in 1887.

Annexationist calls for support by the United States government brought US Navy sailors and marines into Honolulu, the capital city of Hawaii, at the behest of American diplomats to intimidate the weak monarchical government, which then fell almost immediately with only a single shot fired against local police by the annexationists. Fearful of their ability to hold possession of the country, however, the Hawaiian-American rebels called on the American government to support them directly and the American ambassador quickly established a temporary protectorate over the islands backed by the power of the United States Navy. A Provisional Government made up of the Caucasian settlers in the country was established under the American protectorate and following a petition to Washington an annexation treaty was quickly introduced into the US Congress.

The annexation treaty would have likely passed immediately but for the fact that the Republican Administration that favoured immediate annexation, and whose diplomatic and military representatives in Honolulu had aided the American-descended rebels in their revolt, had just been ousted in a federal election, and the US government was in the process of turning over to the Democratic Party. The incoming Democratic President, Grover Cleveland, spurned the annexation of such a far-off country and he believed that American officials had acted improperly towards a benign and friendly state. Cleveland terminated the American protectorate over the islands soon after taking office.

Politically, the Hawaiian Islands languished for some time afterwards, and although the white minority kept their political control they did not do so through democratic means. Rather, the new Republic of Hawaii they created was perhaps more similar to one of the colonialist and apartheid states established in the mid-twentieth century in Southern Rhodesia and South Africa. But the world was becoming smaller owing to new forms of transportation, such as steamships, and it was well known that Hawaii was strategically located in the North Pacific. From Hawaii, much of the Pacific Ocean could be dominated by naval power. That fact became clear to policy-makers in the United States when the US once again found itself under a Republican Administration in the final years of the nineteenth century. The new Republican Administration of President William McKinley began preparing to establish an external US empire.

The United States embarked on a war of conquest against Spain and the US succeeded in taking possession of Spain's colonies in the Caribbean Sea, as well as the Philippine Islands in the western Pacific. The acquisition of the Hawaiian Islands now became crucial to the expansion of American power in the Pacific but to avoid continuing domestic opposition to Hawaii's annexation in the US Congress the McKinley Administration chose to forego a formal annexation treaty between sovereign states in favour of taking possession of Hawaii under a simple congressional resolution that accepted a transfer of sovereignty from the ruling white minority in the islands. That transfer took place despite the fact that thousands of indigenous Hawaiians signed petitions to the US Congress opposing a US annexation and the loss of their Hawaiian nationality. They were ignored.

As a result, the United States acquired the Hawaiian Islands in 1898. Hawaii became an American territory in 1900, and once American sovereignty was established the democratic rights of the native population in territorial matters were restored by the government in Washington. After a long period as an American external territory, punctuated by a Japanese aerial attack in World War II, Hawaii was admitted to the American union in 1959, as the only state that does not form part of the land mass of the North American continent. In the following half-century, the city of Honolulu was transformed from a picturesque South Seas seaport into a concrete metropolis, perhaps now more reminiscent of southern California than anything written into the romanticised South Seas literature. Hawaii today is a top tourist destination for people from North America and from East Asia. It is also the main US base for the projection of US military power throughout the Pacific Ocean.

2

OAHU - THE ISLAND OF HISTORY

In the autumn of 1987, when I found myself in San Francisco, California for a professional conference, I seized upon the proximity of San Francisco to the Hawaiian Islands to plan a visit to those famous Pacific islands.

After a five-hour flight from San Francisco on one of Continental Airline's DC-10 jumbo jets, I arrived in Hawaii as a *malahini*, a first-time tourist visitor to Hawaii. After breezing through the open-style Honolulu airport I found myself standing under swaying palms amidst a darkening sky. I collected my luggage and flagged a taxi to take me to Waikiki, the centre of Honolulu tourism, where I had reservations at the Hobron Hotel. The Hobron was not a luxury hotel on the oceanfront by any means but a rather nondescript skyscraper hotel erected on one of the interior streets of Waikiki not far from the Ali Wai Canal that marks the boundary of Waikiki. I took a very basic upper-floor room at the Hobron that gave me a small partial view of the Pacific Ocean (if I carefully looked for it) between two other hotels that were built much closer to the ocean. Although a high-rise hotel, the rooms at the Hobron did not come with a balcony and my room was rather small and basic. Still, the hotel did have a small outdoor swimming pool and an adequate commissary. Simple but sufficient, and the price was reasonable, especially since October was a shoulder season in the Hawaiian tourist trade, and all of the hotels offered somewhat lower prices at that time of the year. Despite the simplicity of my hotel accommodations they were comfortable enough. Before departing Canada, Carol, my travel agent back in Saskatoon, Saskatchewan, had told me; "You won't be spending much time in your hotel room anyway, so you don't need to spend on a lavish room." And, as it turned out, she was quite right about that.[1]

Honolulu's Hobron Hotel in 1987

After settling in at the hotel off I went to explore Waikiki and Honolulu. Waikiki was, and is, the centre of Honolulu's attractions. Once a retreat for Hawaiian royalty it became an urban concrete jungle in the latter part of the twentieth century but it hosts one of the most well-known beaches in the world. The beach at Waikiki is a long thin strip extending more than three kilometres along the oceanfront. I generally frequented the beach near Fort DeRussey because of its proximity to the Hobron. Fort DeRussey is a military reservation at Waikiki that once harboured a coastal artillery battery. However, the guns of the battery at Fort DeRussey were only fired once because the sound of the artillery blew out the windows of the nearby hotels! The coastal battery was long gone when I visited but the site was still maintained as a rest and recreation destination for US military personnel. However, the beach at Fort DeRussey was open to the public, as are all beaches in the Hawaiian islands, thanks to legislation passed in the days of the monarchy, and the Fort DeRussey beach was much broader than most of the beaches at Waikiki. Its soft white sands were very inviting. Out in Waikiki Bay at that time there were three anchored floating pads that were within swimming distance of the beach. One of the floating pads was used by a parasailing company, Aloha Parasail, and I used their service to go parasailing high above the waters of Honolulu, held aloft by a large red, white, and blue parasail that stretched out behind me while I was being pulled by a long tether attached to a motorboat. Strapped into my parasail by a secure harness I went up higher than 250 metres in the air for an exhilarating look at Waikiki and Honolulu harbour. Ships below me looked like bathtub toys, and the people swimming in the water were tiny specks while above me was the blue sky. It felt like, and in fact it was, a form of flying without wings.

The beaches at Waikiki looking past Fort DeRussey Beach towards the iconic Diamond Head in 1987

Parasailing and swimming were not the only water-sport attractions at Waikiki. Obviously, Hawaii is most famous for surfing but the wave action in October was quite gentle and I skipped trying my luck at surfing. I did try water-skiing but I have to admit that I was an abject failure at it, much to my dismay. I did better in a short scuba dive at Hanauma Bay, where I saw all manner of tropical fish, including a small barracuda.

Back on the beach, the sunsets at Waikiki were gorgeous, with the sky turning shades of red and then blue and purple as the sun slowly sank below the horizon. In the late afternoon, people would often congregate on the beach to watch these spectacular Pacific sunsets. Then, very quickly, the sky would turn dark and the stars would begin to shine. After sunset, there was very little twilight in the tropics, and the darkness swiftly enveloped Waikiki. But it was magical to watch the sunsets unfolding as the rolling waves continuously challenged the sands of the beach with a monotonous roar.

Sunset at the Fort DeRussey Beach, Waikiki, 1987

The beautiful beach at Waikiki attracts people from all over the world and I met all manner of visiting tourists. I made the acquaintance of a young woman visiting from Switzerland and another from Great Britain, whose husband had been lucky enough to score a conference in Hawaii. From the South Pacific, there was a nice couple from New Zealand, who emphasized to me their country's Commonwealth connections with Canada, and finally, I met a band of exuberant young Australian men who were in Hawaii displaying all of the high spirits of youth.

Next to the beach activities, tourist shopping was the daytime terrestrial pastime in Waikiki, and Waikiki did have some interesting shops but mostly it was a place for cheap foreign-made souvenirs. However, I found one store that dealt in Pacific Island carvings and collectibles and I was able to acquire a finely carved talking stick from the Trobriand Islands of Papua New Guinea, and a well-made tapa cloth from Tonga for a reasonable price. The proprietors told me that Hawaiian traditional crafts were somewhat of a lost art resulting from the disappearance of the indigenous population, although handmade shell necklaces from the forbidden island of Niihau, where a remnant indigenous population still existed, were authentic examples of Hawaiian craftsmanship. Those Niihau shell necklaces were highly collectible, and therefore expensive, selling for very high prices both then and now.

Eating establishments in Waikiki ran the gamut of American cosmopolitanism but most eateries were largely geared to the mass-market tourism crowd and included American fast-food outlets and buffet-style restaurants for the budget-conscious tourist. The major resorts supplied a more upscale dining experience, often giving diners very attractive views of the ocean as they dined. But there was also a seedier side to Waikiki, in the form of down-market bars and strip joints. In the late evening, streetwalkers could be seen patrolling darkened street corners darting out from the shadows, accosting men to sample their favours.

I finished my exploration of Waikiki's ersatz Polynesia with an evening harbour cruise, perhaps more commonly and accurately described as a booze cruise, replete with drinks, dinner, and hula dancing performed by young women. The hula dancing was mainly of the tourist variety, as the women swayed, swished, and swirled in plasticized ti-leaf skirts but at least one of the solo dancers did appear to have had some degree of traditional training, and offered a more authentic form of hula, with the traditional slow and graceful movements of the arms and hands.

Harbour Cruise Hula, 1987

Having satiated any desire for pure tourist escapism at Waikiki I decided to explore the farther reaches of Honolulu and Oahu, searching for a missing Polynesian authenticity, and for that purpose I needed a car. So I made arrangements with a car rental agency and Janet, a vivacious Chinese-Hawaiian woman, collected me at the Hobron and took me to a pick-up location for the Ford Mustang convertible rental that I had selected. As we drove along I asked her about conditions in the islands and uppermost in her mind was the encroaching influence of Japan in

Hawaii. In the late Eighties, the Japanese economy was booming, going from strength to strength, and giving rise to American fears that Japan might even supplant the United States as the world's economic powerhouse.

Hawaii, of course, is in much closer proximity to Japan than the American mainland and the islands were a major tourist draw for holidaying Japanese. Waikiki itself was filled with Japanese tourists on package tours to Hawaii, giving rise to parallel tourism with Japanese tourists on group holidays, and western tourists on individualized holidays. But it was not Japanese tourism that created the local concern about Japanese influence in the islands. Rather it was the economic impact of the Japanese investment flowing into the islands that caused the growing concern. Janet explained that wealthy Japanese were coming to Hawaii to buy real estate and they were willing to pay substantial sums beyond the local market-price levels. In some cases, she said, Japanese purchasers had been known to drive around Oahu stopping at a house that they liked and simply knocking on the door and offering the owner an extravagant amount of cash for their house. Such spontaneous offers were sometimes attractive enough to tempt a homeowner into selling despite having no previous intention to sell their home. The effects on the price levels in the local real estate market due to the infusion of offshore money had quickly become apparent, and the *kama'aina*, a Hawaiian word meaning the residents of the island, worried that they might soon be priced out of their own housing market.

Oahu, in fact, was always the epicentre for the historic competition between the United States and Japan for the Hawaiian Islands. It was the Americans who had made the initial inroads into the island chain through the influence of the Congregationalist missionaries who came to the islands in the early nineteenth century, and whose descendants subsequently dominated the economy of the Kingdom of Hawaii before coalescing politically around the effort to overthrow the kingdom and annex the country to the United States. In the meantime, the need for labour to work the sugar cane plantations had forced the foreign planters to persuade the King of Hawaii to permit extensive immigration, and it was from Japan that Hawaii obtained the majority of its immigrants in the nineteenth century. The demographic growth of the Japanese community in Hawaii eventually became a source of anxiety to the planters, who feared that they might lose their grip on the country, and that prompted them to pursue American annexation in order to forestall the day when the size of the ethnic Japanese community would translate into Japanese political influence in Hawaiian affairs. In the years after the monarchy was overthrown, and while waiting for Uncle Sam to absorb the country into the United States, the white-minority government of the Republic of Hawaii often found itself in serious diplomatic difficulties with Japan because of Hawaii's attempts to restrict both Japanese immigration as well as the civil rights of Japanese settlers in the islands.

Even after the US annexation of the islands efforts were made by the dominant white minority to politically suppress the Japanese-Hawaiian community by denying American citizenship to those Japanese settlers who had been born abroad, although it was not possible for the annexationists to prevent any Japanese children born in Hawaii from obtaining American nationality. By the Thirties, the Japanese community was beginning to have an electoral impact as Japanese-Americans born in Hawaii after annexation reached their age of majority and began to vote in territorial elections. The white community that had ruled Hawaii since the overthrow of the monarchy began to worry that its control over the politics of the islands might be imperiled as the demographics of the electorate changed over time. However, the most significant racial incident during the interwar years involved the murder of an indigenous Hawaiian by white vigilantes whose crime was subsequently excused by the US military authorities. Real political change would be delayed by World War Two and would not come to the islands until the Fifties.

My circumvention of Oahu took me around the island and true to its reputation I discerned the love of Hawaiian residents for flowers. Many different varieties are cultivated in the Hawaiian islands including gardenia, plumeria or frangipani, jasmine, which is here called *pikake*, meaning peacock, as a tribute by Hawaiians to their beloved Princess Kaiulani, who loved both the scent of jasmine and the beauty of her pet peacocks. Bougainvillea sprouts everywhere on the island, and there are various colourful gingers and orchids, the *ilima*, and the scarlet *ohia lehaa* that grows on the slopes of the great mountains on Hawaii Island. Green Pandanus and the Ti plant are used in traditional Hawaiian hula skirts and decorations.

All of these flowers but especially orchids find their way into flower leis, the decorative neckpieces worn by the men and women of Hawaii on special occasions. Once leis were given to every tourist on their arrival in the islands by ship but as the number of tourists grew into the millions with aeroplane travel that custom became more honoured in the breach than the observance, although also I found that custom had expanded beyond Hawaii to other Pacific islands. When I visited Hawaii tourists bought cheap plastic imitations of the real leis as a souvenir. But Hawaiian residents still give each other real flower leis on important occasions, such as christenings, birthdays, and graduations, as an expression of the Hawaiian spirit of *aloha*. The Hawaiian word *aloha*, which is often heard on Oahu, is a greeting that has multiple meanings and depending on the circumstances can include the breath of life, love, friendship, welcome, or a farewell. The first day of May is still marked in the state as Hawaiian Lei Day to honour the historic custom.

I drove to the North Shore of Oahu, and to the coast that is famous the world over for the gigantic surfing waves that rush ashore here. The most important surfing competitions on Oahu take place on the North Shore, which is a mecca for professional surfers and talented amateurs alike. However, I was visiting in October, which was just a bit too early for any of the major competitions. The best time for surfing on the North Shore in Oahu is said to be between November and February and that is when the important surfing competitions are held.

Along the North Shore, I stopped my car to pick up a hitchhiker, a young blond-haired man, and as I drove along the coastal road with the convertible top down I chatted with him about life in Hawaii. His ambitions, I learned, were bound up with his passion for surfing, which he did as often as he could, to the exclusion of much else. But when I asked him how he earned his living he told me that he supported himself by selling narcotics! Another marker of the Americanization of the islands!

As I drove through Oahu I perceived that something was missing – the Hawaiians. The indigenous people of the islands and their unique culture had disappeared, and have been replaced by a version of American culture that is overlain with a thin Polynesian veneer. In truth, the indigenous Polynesian inhabitants of these islands were largely gone. By the early 2020s, the US Census Bureau would report that the worldwide population of pure Hawaiians was about 5000 people, although there are still many who can claim a partial indigenous descent. The Hawaiians who did not die out as a result of the waves of infectious diseases that spread across the islands in the nineteenth century saw their traditions and culture suppressed and absorbed into the American melting pot. Of the remaining people who claim some form of indigenous Hawaiian ancestry many had fallen behind economically, and socially, by the time of my visit in 1987, and they remained among the poorest demographic groups in the islands. Successive waves of immigration and cultural contact from the Occident and the Orient had devastated the original indigenous culture. Today, Hawaii is an amalgamation of cultures, principally American and East Asian, although the original Polynesian culture of Hawaii is memorialized in a number of ways by the current inhabitants of the state.

There was one place on Oahu that provided a verisimilitude experience of the once-dominant Polynesian culture of Hawaii and that was at the Polynesian Cultural Centre located in the north-western part of the island, where it is associated with the Mormon university on Oahu. There, students attending the university from other Pacific islands display their island cultures in a somewhat Disneyesque environment. Separate villages tell the individual stories of each island, and expressive dance performances display something of the cultures of various Polynesian and Melanesian islands. It serves as a replacement experience of the indigenous Hawaiian culture for the modern tourist.

Traditional dancing at the Polynesian Cultural Centre on Oahu, 1987

To learn more about the Hawaiian Polynesian culture that once prevailed across these islands I went to the Bishop Museum, one of the great anthropological repositories of the South Seas islands. There, a vast collection of artifacts and written observations of the Hawaiian Islands obtained in the heyday of the Hawaiian race is preserved for posterity. The main hall with its extensive use of native woods was impressive and architecturally attractive. The Bishop is the grand repository of Hawaiiana – especially of the royal period that dominated the nineteenth century when the Kingdom of Hawaii was an independent nation that was diplomatically recognized by many countries. During that period Hawaii's Polynesian rulers were received as equals in the royal courts of Europe, and South America, and in the great halls of government in Washington.

After learning of Hawaii's prominent role in nineteenth-century Pacific history at the museum I resolved to visit Iolani Palace in downtown Honolulu, once the seat of the Hawaiian monarchy, and the only royal palace on US soil. Iolani Palace was built in the 1880s as a royal residence for the Sovereigns of Hawaii in an architectural style that is now known as American Florentine, which is a mixture of the Italian Renaissance style and Hawaiian elements. The palace is the only extant representative example of this American Florentine style, although there are some other buildings in tropical settings elsewhere in the world that were constructed during the same time period, and which are architecturally similar.

The palace is built of brick, faced with concrete, and has six towers, four at each corner and two centre towers in the front and back of the building. Two lanais or verandas, one on top of the other, wrap around most of the building. Iolani Palace was a very advanced building when it was constructed and it possessed indoor plumbing, electrical wiring, and telephones, all of which were unusual and remarkable when the building was constructed. The Hawaiian monarch's home actually had electricity installed before the White House in Washington was wired for electricity. Beautiful woods were used throughout the interior, especially on the formal staircase leading from the main floor to the upper floor.

Iolani Palace, Honolulu, Hawaii

I went on a guided tour of the palace with a group of tourists, mostly American. We had to remove our shoes and put on soft foot covers to protect the wooden floors. Our tour guide explained the highlights of the palace building itself as well as giving us a history lesson about the latter days of the Hawaiian monarchy, the revolt by the white settlers, and the American annexation of the islands. On January 17, 1893, white settlers, mostly American, seized the government buildings in Honolulu while US Marines and US Navy sailors landed from a US warship and established a military camp, and commenced street patrols in the Hawaiian capital. The white settlers issued a proclamation abolishing the monarchy and establishing a provisional government. In the face of the obvious US support for the rebellion the weak royal Hawaiian government collapsed, and the following day the royal standard was lowered from atop Iolani Palace as the now ex-Queen Liliuokalani left her palace and returned to her nearby private residence at Washington Place. Some of the American tourists were surprised by the history they heard, and one American man sputtered "What – you mean we did it again!" when our guide referred to the American-led coup and the landing of US forces in Honolulu. He was thinking, perhaps, of a well-documented practice of the United States in arranging for the replacement of foreign governments that it dislikes.

Actually, Hawaii was the first example of American diplomats and military officials becoming involved in a coup in a foreign state, and Hawaii was the only such example that actually led to a permanent American annexation of the foreign state. A poignant moment in our tour came in the former royal throne room when our guide softly said "It was in this room that Queen Liliuokalani learned that she had been overthrown."

In the following years, the new white minority government of the Republic of Hawaii did its best to extinguish the symbols of the monarchy. The palace furnishings were sold at auction (although some were apparently just taken from the palace as spoils). In his book *Rascals in Paradise*, the author James Michener recounts the tale of how one of the Provisional Government's soldiers discovered the King's Crown in the basement of Iolani Palace and pried out its precious jewels, which he then fenced on mainland America. Today, the King's Crown on display in the Iolani Palace holds fake jewels. However, the companion Queen Consort's Crown is authentic but only because the widow of King Kalakaua wisely took her Crown with her to her private residence when she left Iolani Palace after the death of the King.

In the aftermath of the overthrow of the monarchy, Iolani Palace was renamed the Executive Building, and it was converted into the legislative and executive offices of the Republic of Hawaii and, after annexation, of the Territory of Hawaii. It was only in the late Sixties, after the construction of a new Capitol Building, that the State of Hawaii vacated the building and turned it over to a non-profit group that sought to restore the building as a memorial to the Hawaiian monarchy. Since then great efforts have been made to restore and preserve Iolani Palace and its success as a preservation project marks the determination of the remaining Hawaiians and *kama'aina* to remember the past of these islands.

Outside on the grounds as I emerged from the palace a real relic of the days of the monarchy assembled on the Royal Coronation Pavilion adjacent to the palace. It was the Royal Hawaiian Band, which was created by order of King Kamehameha V in the mid-nineteenth century and preserved in existence since that time, despite the many political upheavals that had occurred in these islands. Dressed in their nineteenth-century Ruritanian-style uniforms, replete with pith helmets, the band proceeded to play a selection of marches as well as a number of traditional Hawaiian songs, including the most famous Hawaiian song, *Aloha Oe*, a song which was composed by Queen Liliuokalani herself.

The palace tour guide had mentioned that a statue of Queen Liliuokalani had been erected outside, between the old palace and the much newer State Legislature, and she said that every day a small bouquet of flowers was placed in the Queen's outstretched hand by an unknown person. After touring the palace I went to see the statue of the Queen and, sure enough, there were flowers in her hand – positive proof of a continuing loyalty by some in these islands to their ancient nobility.

The statue of Queen Liliuokalani in Honolulu shows the flowers placed in her hand by a contemporary loyalist

The nearby Capitol Building was of very modern construction and I wandered into it and was able to join a group tour that highlighted the architectural features of the building as a whole, and the legislative chamber in particular. The architectural style of the building is a variation of the modern Bauhaus style of architecture but it is unique among American legislative buildings in that it features an open-air design that allows the climate to penetrate into the agora or central space of the building. The legislative chambers on each side of the building are also unique in possessing a cone-shaped design that is intended to symbolize the volcanic origins of the Hawaiian Islands. The Capitol Building also houses the offices of the Governor of the State of Hawaii and it is said that in the Sixties the former US First Lady, Jacqueline Kennedy, who was spending time in Hawaii to heal following the assassination of her husband, the President, had influenced the interior design of the portion of the building that houses the office of the Governor of Hawaii.

During the tour of the legislative chamber, our guide drew our attention to the Hawaiian State Flag which is exactly the same as the once-national flag of the Kingdom of Hawaii. The state flag portrays the British Union Jack in the canton and on the fly contains eight large alternating blue and white stripes that symbolize the eight major islands of the Hawaiian chain. It is an unusual example of American vexillology (the study of flags) because of its use of the British Union Jack, the symbol of what Americans rebelled against in 1776. Its use in Hawaii reflects the historical friendship between the British Empire and the Kingdom of Hawaii, and its retention after the annexation was intentional to soften the indigenous opposition to the loss of Hawaiian independence.

The Hawaiian Flag billows above Washington Place, the former residence of Queen Liliuokalani and the Governors of Hawaii

Visiting Iolani Palace certainly wetted my curiosity about nineteenth-century Hawaii, and its monarchical era. Much of the history of Iolani Palace was concerned with the last Hawaiian dynasty. I was fascinated by the story of how King Kalakaua had striven mightily to obtain and hold the support of western countries for the continued independence of Hawaii through his trips abroad to the United States and by his round-the-world tour, the first by a head of state, which brought foreign attention to Hawaii. But foreign, primarily American, settlers wrested control of the economy from native hands leading to a revolt in 1887 that stripped the King of most of his powers. The King died in 1891, without a son or daughter to succeed him, and his sister, Princess Liliuokalani, succeeded him on the Hawaiian throne only for a short period before the final settler revolt.

I also learned about a charming young Hawaiian princess named Kaiulani, who was a student in England when the *coup d'etat* occurred in Honolulu that brought the monarchy to an end. She hastened to America in spite of her youth and unsuccessfully sought to convince Americans to overturn the results of the coup and to support both the Hawaiian monarchy and the independence of Hawaii. Although President Grover Cleveland did delay the American annexation of Hawaii for the length of his four-year term of office the subsequent US war with Spain, and the seizure of the Spanish Philippines, made the acquisition of Hawaii imperative for the projection of American naval power in the Pacific, and the fate of independent Hawaii was sealed. The formal annexation of the Hawaiian Islands by the United States occurred on August 12, 1898, and Hawaii's star entered the American firmament.

To learn something about the earlier Kamehameha dynasty, I visited Queen Emma's Summer Palace in the Nuuanu Valley, also known by its Hawaiian name Hanaiakamalama. Queen Emma's Summer Palace was constructed in 1857 as a summer retreat for the then-ruler of Hawaii, King Kamehameha IV, and his Queen of mixed Hawaiian and Scottish ancestry, Emma Rooke. It is really more of a villa than a palace. Here, the royal couple enjoyed the cooling breezes of the Nuuanu Valley during the hot Hawaiian summers. Fourth in the line of Hawaiian monarchs, King Kamehameha IV was an intelligent monarch of whom much was expected.

Emma Rooke was the daughter of a prominent Hawaiian *ali'i* and a Scottish father and in light of her genealogy, she was also considered to be eligible to hold the Hawaiian throne in her own right. Together, this Hawaiian power couple ruled a country that was emerging onto the world stage and they sought to promote good relations with the major western powers, especially Great Britain. They brought the Anglican Church to Hawaii and Queen Emma, with the support of her husband, was the founder and initial benefactress of the Queen's Hospital, an institution that is still a major medical facility in modern Hawaii. But there was a worm in the spirit of the King. He was a high-strung and somewhat mercurial man, and when his Queen formed too close an attachment to the King's aide-de-camp King Kamehameha IV shot his aide, giving him a wound that would eventually claim the poor man's life. Finally convinced by his wife and others that he had been deceived by scurrilous rumors, and filled with remorse, the King contemplated abdication in order to subject himself to the law but he was dissuaded from doing so by his advisors. Then, Prince Albert Edward, the only child of the royal couple became ill after the King bathed his son in cold water and his parents watched helplessly as the young boy died of appendicitis, or possibly meningitis. Deep in grief and guilt for the deaths of both his aide and his son the young King withdrew from public affairs. His health declined precipitously and he died a short time later at the early age of twenty-nine.

Hanaiakamalama preserves the memory, and many of the artifacts of this royal couple and especially of Queen Emma, who remained in correspondence with Great Britain's Queen Victoria for many years after the deaths of her son and her husband. Queen Emma also went on to play a prominent role in Hawaiian politics, and in Hawaiian society, until her death in the 1880s.

The history I learned confirmed what my eyes could tell – that the ancient Hawaiian race had largely disappeared from Hawaii, eliminated by waves of diseases, or merged into the populations of Americans, Europeans, and Asians who had flocked to these shores. At the time of my visit in 1987, Hawaii possessed an Asian majority and a white minority- the only American state exhibiting such a unique demographic profile. Chinese and Japanese temples abounded and one day I witnessed a local Buddhist priest offering blessings to celebrate the opening day of a new store in Waikiki.

Despite all the changes that had occurred in these islands, however, a Hawaiian cultural renaissance was underway and those residents who could still claim some Hawaiian ancestry were taking steps to recover what they could of their former Hawaiian culture. Much of their traditional knowledge had been lost but fortunately, there were still places in the wide Pacific that preserved the old island ways. From the South Pacific came the knowledge of how to make tapa cloth, or *Kapa* as it is called in Hawaii, and from the Caroline Islanders in Micronesia native Hawaiians learned once again how to navigate at sea by the stars and the currents of the ocean. A turning point for the remaining Hawaiians came with the construction and sailing to Tahiti in 1976 of a traditional Hawaiian ocean-going canoe, the 18-metre *Hokule'a*. The crew of the *Hokule'a* navigated their way to Tahiti in the Society Islands in the same way that their forefathers had once found their way from the Society Islands to Hawaii – by the stars and the ocean currents. The building and sailing of an ocean-going canoe in Hawaii sparked a renewal of Hawaiian pride, and it led to a recovery of some traditional practices.

Nevertheless, I still found Hawaii to be a thoroughly American society, though a delightful and beautiful place that had preserved a memory of the original inhabitants, and of some of their customs. Towards the end of my stay on Oahu, I went down to the Honolulu harbour front. While sitting there I struck up a conversation with an older man of apparent Hawaiian descent and I expressed my deep admiration for the beauty of his homeland. He asked me where I was from and I told him that I was from Canada. He thanked me for visiting and he expressed the hope that I would enjoy my visit but he also stated that he and others of like mind felt that too many visitors were coming to these islands. He had seen old ways give way to a modern way of living – one that was often wasteful and out of sync with the ancient Hawaiian spirit. I thanked him for telling me that and I said that I hoped that his islands would always remain a beautiful place for the people who cherished them.

NOTES

[1] Much later the Hobron hotel became a condominium.

3

PEARL HARBOUR – THE BIRTH OF A WORLD POWER

It is virtually *de rigeur* for any visitor from North America of my generation to visit Pearl Harbour, the great American naval fortress in the Hawaiian Islands, and the scene of the opening battle of the Pacific war in World War II. The surprise Japanese aerial attack on the US Pacific Fleet on Sunday morning, December 7, 1941, was a global event that transformed America and the world by turning the United States from a weak isolationist giant into a global economic and military colossus. The entry of the United States into World War Two following the Battle of Pearl Harbour guaranteed the ultimate victory of the Allied and Associated Powers against the fascist Axis Powers.

The United States and Japan initially had favourable relations with each other in the years after the US Navy's Commodore Matthew Perry forced open the gates of Japan in the 1850s. Good relations continued after the Meiji Restoration in 1867, which marked the beginning of the stratospheric rise of Japan as an industrial and military power. But America looked on anxiously as Japan became progressively more dominant in East Asia, first by defeating China in 1895 and then defeating Russia in 1905 before absorbing Korea into the growing Japanese Empire in 1910.

In 1914, upon the outbreak of World War One Japan sided with the allied powers and seized the German colony at Tsingtao in China after a brief but hard-fought campaign. More important from the American perspective was the naval campaign mounted by Japan to take over the German-owned Mariana Islands, Caroline Islands, and the Marshall Islands in Micronesia, giving Japan a vast new oceanic empire in the western Pacific. American strategists were alarmed because the Japanese-owned islands now lay astride the direct line of communications between the US-owned Hawaiian Islands and the US-owned Philippine Islands. Even before World War

One Japan increasingly threatened China's control over Manchuria, and in the interwar years, Japan succeeded in detaching Manchuria from China and establishing a Japanese puppet regime there.

The United States increasingly found disfavour with the policies of the Japanese government, as Japan became increasingly militaristic and fascistic with the passage of time. The US government took particular umbrage at Japanese efforts to conquer and dominate vast tracts of China and the stage was set for a titanic collision between Japan and America. America's decision to cut off vital oil supplies to Japan in 1941, as a result of Japan's incursion into French Indochina (with the connivance of the pro-Axis Vichy French government), brought their competition in the Pacific to a head, and the cataclysm broke with the Battle of Pearl Harbour.

The Japanese operation against the US Pacific fleet at Pearl Harbour was a masterful demonstration of Japanese naval planning and risk-taking. Planning was commenced in great secrecy by Admiral Isoroku Yamamoto, the commander of Japan's Combined Fleet. Unlike other leading military figures in Japan Yamamoto was no war-monger because he knew the strength of the United States, having served as Japan's naval attaché at the Japanese embassy in Washington. But when the Japanese government decided to conquer the European empires in south-east Asia from the French, British, and Dutch, Yamamoto knew that the US Pacific Fleet was the only military force that could potentially stop Japan's expansionist plans. Japan would have to damage, or destroy, the US fleet in order to embark upon its plan of conquest in the Pacific.

Admiral Yamamoto concocted a daring plan to send a battle fleet including six of Japan's aircraft carriers through the rarely traversed waters north of Hawaii to begin hostilities between Japan and America by launching a surprise attack on the Pearl Harbour naval base, and hope to catch the Americans napping. It was a high-risk gamble because the Japanese fleet had to perfect refuelling at sea and it also had to design a method of torpedo attack that would prevent any Japanese torpedoes launched from torpedo bombers from getting stuck in the mud of Pearl Harbour. Above all, complete secrecy was required because if the Japanese battle squadron was detected during its voyage in the seas north of the Hawaiian Islands it would be extremely vulnerable, and it might be destroyed by the Americans before it could threaten Pearl Harbour.

Although the Japanese plotted their attack in great secrecy the United States knew that the Japanese were up to something. The United States military had an excellent knowledge of Japanese warmongering because it had broken the Japanese diplomatic codes, and both the US army and the US navy were routinely intercepting military and diplomatic communications through a special crypto-analytical project known as 'Magic'. One intercept between Tokyo and the Japanese consulate in Honolulu was particularly telling. In the so-called 'bomb-plot' message Tokyo tasked its consulate in Honolulu to specify where each type of US warship (e.g. battleships, aircraft carriers, cruisers etc.) berthed within Pearl Harbour. Such information would only be required as part of the detailed planning for an attack, and that should have tipped off the American admirals that an attack was coming. But neither that message nor others jarred the complacency of important political and military officials in Washington and Pearl Harbour and none of the responsible officials perceived the looming danger of a Japanese air raid on Hawaii.

The failure to identify and forestall a Japanese attack was not a failure of American military intelligence gathering but rather a failure to understand the implications of the intelligence that was gathered. US officials, both military and civilian, who read the Japanese messages often asked themselves; 'What does this message mean?' But they would have been better served by asking themselves; 'what is the worst thing 'that this message could mean?' There was also a racial bias at play in the analysis of Japanese communications because many American officials could not comprehend that an Oriental race could mount a sophisticated assault on America's mid-ocean fortress. Even when a mobile radar station on Oahu reported the approach to the island of a mass of aircraft on the morning of the attack there was still no recognition of possible danger, and the radar contact was merely attributed to an expected arrival of B-17 bombers passing through Hawaii on their way to the Philippines.

On November 26, 1941, a Japanese strike force consisting of six aircraft carriers, two battleships, and a bevvy of cruisers, destroyers, support ships, and submarines sailed from Hitokappu Bay in Japan under the command of Vice-Admiral Chuichi Nagumo to attack the US fleet at Pearl Harbour, Hawaii. Nagumo was a surface fleet specialist rather than an air combat officer but he relied on several air specialists to plan and carry out the aerial attack. A key officer in the planning of the Pearl Harbour attack was Commander Minoru Genda, who worked closely with Admiral Yamamoto, and with Commander Mitsuo Fuchida, the pilot who led the actual aerial attack on the US fleet. Together, they designed the plan that was put into effect.

At 8 AM on Sunday, December 7, 1941, a quiet weekend morning at Pearl Harbour, Japanese torpedo bombers, dive bombers, and high-altitude bombers arrived over the US Pacific fleet. The US fleet, under the command of Admiral Husband Kimmel, was riding quietly at anchor and was completely unsuspecting. Within minutes there was utter chaos as Japanese bombs and torpedoes tore into the US fleet. Admiral Kimmel stood on the hillside front lawn of his neighbour's home and watched stunned as the Japanese Navy annihilated the bulk of his fleet. One by one the battleships that formed the backbone of the fleet went down. The *USS Arizona* blew up and was destroyed. The *USS Oklahoma* capsized and sank. Both ships were total losses. Two other battleships sank into the mud and one other battleship was beached while three others were left in a damaged state. A total of twenty-one US warships were sunk or damaged during the attack but in a stroke of luck, all of the US Pacific Fleet's aircraft carriers were away from

Pearl Harbour at the time of the attack, the *USS Hornet* and the *USS Enterprise* at sea, and the *USS Yorktown* refitting at the US west coast. They remained available for future offensive operations.

A poignant reminder of the battle that day is the *USS Arizona Memorial*, which lies over the shattered hulk of the destroyed battleship. The *Arizona* was still leaking oil from the attack when I visited the site in 1987. Built atop the sunken hulk of the battleship, the *Arizona Memorial* pavilion marks this site as a naval graveyard, the final place of entombment for those American sailors who went down with their ship on the morning of the attack. It was a place of hushed sadness among those Americans who I saw visiting the memorial in 1987, but there was also an enduring respect for those who had fought and died there. The names of the dead are inscribed on a far wall of the pavilion and as I read the names I suddenly came across my own surname. One of the American sailors who had served and died on *Arizona* was also named Noonan. In reading our name I recognized a symbolic link between myself and the men who had served and died to preserve the freedoms and values of western societies.

The USS Arizona Memorial Pavilion at Pearl Harbour, 1987

In the days that followed the Battle of Pearl Harbour the Japanese attacked widely across the Pacific, commencing campaigns of conquest against the US-owned Philippines and Wake Island, as well as landing forces to occupy British-owned Malaya, and forcing independent Thailand to enter into an alliance with Japan. The Japanese forces already in China besieged the British colony of Hong Kong on the China coast, where a British and Canadian garrison was heavily outnumbered and forced to surrender at Christmastime. The British navy suffered an immense blow when Japanese aviators found and sank the modern battleship *HMS Prince of Wales*, and the older battlecruiser *HMS Repulse* off the Malayan coast despite the fact that both battleships were armed with anti-aircraft guns, and were protected by an escorting screen of destroyers with their own anti-aircraft armament. The destruction of the two British capital ships and the earlier battleship losses by the US Navy at Pearl Harbour marked the end of the supremacy of the battleship on the world's oceans and the rise of the aircraft carrier as the capital ship of modern navies.

A portion of the wreckage of the Arizona at Pearl Harbour. The Japanese attack on the US Fleet marked the end of the primacy of the battleship on the world's oceans

After leaving the *Arizona Memorial* pavilion I visited the nearby interpretative centre that told the story of the Battle of Pearl Harbour, and the subsequent naval campaigns in the Central Pacific that were directed by US Admiral Chester Nimitz from Pearl Harbour. Although defeated at Pearl Harbour, the US Navy would rise swiftly to meet Japan's challenge and together with America's strong manufacturing base, and the availability of the rich resources of North America, the United States would create a military juggernaut that would eventually overwhelm and defeat the Japanese Empire.

I finished my visit to Pearl Harbour with a tour of the *USS Bowfin* – a World War Two era submarine that is permanently moored at Pearl Harbour as a memorial to the US Navy's submarine service, which was an essential element of America's ultimately victorious naval campaign against Japan in World War Two.

The USS Bowfin is permanently moored at Pearl Harbour

The Battle of Pearl Harbour remains an inflexion point in modern Pacific and world history. Prior to that battle, the United States was a major power but it was not the dominant power anywhere outside of North America. After the Battle of Pearl Harbour, the United States began a swift rise to world power that would allow it to become the dominant power in the world, and the master of the Pacific Ocean.

KAUAI - HAWAII'S GARDEN PARADISE

During my visit to the Hawaiian Islands in October, 1987, I wanted to visit at least one other island besides Oahu, and I chose Kauai Island as my second island to visit. While outsiders consider all of Hawaii to be a paradise Hawaiians themselves often point to Kauai, the so-called Garden Island, as the quintessential paradisiacal island of the Hawaiian chain.

And so, with anticipation, I boarded an Aloha Airlines flight for the short flight from Honolulu to the town of Lihue on Kauai. At the Lihue Airport, I rented a car and set out to explore this lovely island. Kauai is mostly (but not entirely) encircled by a road and I started from Lihue and then headed towards Hanalei on the north side of the island. Along the way, I passed by the Wailua River, the only river in all of Hawaii that is navigable by boat. As I drove along the coast I journeyed through sleepy coastal towns totally lacking in the urban vibe I found on Oahu. There were no skyscrapers on Kauai, so it proved to be a welcome retreat from the hustle and bustle of Honolulu.

The scenery on Kauai is so spectacular that the island is routinely selected by Hollywood movie directors as the place to film features that have a tropical setting. And that scenery does not disappoint. Kauai is a place of great physical beauty and the landscape that I found was verdant, with lush tropical vegetation sprouting everywhere along the coastal road. Soft rains make visiting this island a green and wet tropical experience and one of the wettest spots on planet Earth is actually found on Kauai. I stopped to admire a small *heiau* ruin near a fork in the road, a partially overgrown reminder of the once vibrant but now evanescent culture of the Polynesians. A rain shower sparkled along the way, and I stopped at a small and lovely deserted beach. I ventured out into the water but the incoming wave action was so strong that it began to smack my body down on the hard sand and tore at me as each strong wave receded. I recognized a danger from the elemental forces of nature and I quickly retreated from the water before I got myself into further trouble.

A beach on Kauai, 1987

Continuing my journey I found that the lushest part of Kauai was where the road proceeded past Hanalei towards the end of the road, where the road reaches the impassable Na Pali Coast. The Na Pali Coast consists of towering mountains and extreme cliffs pierced by waterfalls that tumble down from on high to the ocean below, leaving no room for a road. Visitors who want to see the Na Pali Coast in all its glory must see it offshore in a boat, or aloft in a helicopter. Although it is possible to hike through the mountainous coast to the other side the hike is more than 35 kilometres long through the lush Hawaiian jungle, rugged mountains, and steep ravines, which my time on the island did not permit me to undertake.

A beautiful waterfall on Kauai, 1987

Returning from the Hanalei-Na Pali Coast I headed next towards the south side of the island. The southern side of Kauai is drier and as a result, it is not quite as lush as the road on the Hanalei side although it was still quite beautiful. The road curved towards Waimea beyond Lihue and it took me to one of the most spectacular of all the striking natural features of the Hawaiian Islands, the Waimea Canyon. The Waimea Canyon is a deep gorge 22.5 kilometres long, 1.6 kilometres wide, and more than 1100 metres deep and it is the signature geological feature of the island of Kauai. The canyon formed over the aeons by the collapse of an ancient and now extinct volcano, as well as by the action of water erosion courtesy of the Waimea River, the longest river on the island. The result has been the carving out of the island of a canyon replete with great crags and buttes, all graced with verdant vegetation. During the pre-western contact period, the banks of the Waimea River were the site of an extensive Polynesian agriculture settlement, and there were many villages in the area but now the indigenous Polynesians are gone, and only the memory of them inhabits the banks of this river.

The story of Kauai is unique in comparison to the rest of the island chain. Unlike Hawaii Island, Molokai, Kahoolawe, Maui, Lanai, and Oahu, Kauai and its small neighbour Niihau were never conquered by King Kamehameha I, and they only became part of the larger Kingdom of Hawaii in 1810, following their peaceful submission to the allegiance of King Kamehameha. The peaceful accession of Kauai and Niihau to the Kingdom of Hawaii initially gave the former Kingdom of Kauai and Niihau some minor autonomy within the larger Kingdom of Hawaii for a certain length of time but a subsequently unsuccessful rebellion during the reign of King Kamehameha

II, resulted in Kauai losing whatever autonomy it had preserved, and it was later fully integrated into the Kingdom of Hawaii.

Kauai was lightly populated in comparison to Oahu, and it had more of a rural or small-town feel to it. Perhaps that also reflected the history of the island as a major agricultural area for the growing of sugar cane. (It was at Koloa, on Kauai, in 1835, that the sugar cane industry in Hawaii was founded.) As a result, Kauai seemed to me to be a much more relaxed island environment and a far cry from the urban cacophony of Oahu. The demographic pattern of Kauai was much the same as the rest of the Hawaiian Islands, however, the Asian community was dominant in numbers followed next by the Caucasian community and the original native Hawaiians were now little more than a remnant population on the island. Tourism was ever-present of course, as on all of the major Hawaiian islands, and I spied a significant resort at Princeville as I drove towards Hanalei but otherwise there was much less resort development on Kauai in comparison to Oahu.

The County of Kauai includes both Kauai Island and its unusual neighbouring island, the forbidden island of Niihau, which is visible from Kauai to the northwest. A small population of indigenous Hawaiians still lives on Niihau and Niihau is the only Hawaiian island on which the Hawaiian language remains the ordinary language of the people. Niihau is called the forbidden island because the Sinclair-Robinson family that has owned the island since they purchased it from the King of Hawaii in 1864, for $10,000, has forbidden all non-indigenous Hawaiians from living there, and has even prohibited visitors to the island except by special permission. The family maintains that it had promised the King of Hawaii that the Robinsons would take care of the Hawaiian people who called Niihau home after their purchase of the island, and the family has striven across generations to keep that promise. As a result, Niihau preserved something of a Polynesian culture that could no longer be found anywhere else in the Hawaiian Islands in 1987. Niihau shell necklaces were (and still are) a locally made handicraft, that remain a highly desired Hawaiian collectible.

Although my time on Kauai was short the beauty of this island remained embedded in my mind even after the passing of many years. It well deserved its description as the Garden Island of the Hawaiian archipelago. And so, it was with a certain sense of reluctance that I boarded my return flight on Aloha Airlines to Oahu, where I began my preparations to depart these islands while promising myself that someday I would return.

A coastal scene on Kauai, 1987

PART II

THE ISLANDS OF MICRONESIA

INTRODUCTION TO MICRONESIA

In 1989, two years after my first visit to Hawaii, I decided to visit the remote islands of Micronesia. Ethnographers divide the Pacific islands into three ethnolinguistic groups, the Polynesians in the eastern Pacific, the Melanesians in the southwest Pacific, and the Micronesians in the northwest Pacific. In my youthful days spent poring over maps of the Pacific Ocean the islands of Micronesia seemed very mysterious to me. There seemed to be much less written about them than other Pacific Islands and even the US National Geographic Society in its two Pacific island books, *Isles of the South Pacific,* and *Blue Horizons: Paradise Isles of the Pacific* overlooked them. In large part that was a result of World War Two. The Micronesian islands were the site of many terrible battles in World War II, and after the allied victory the United States received the islands as a United Nations Trust Territory – but with a twist. The Micronesian islands were established as a strategic trust, the only strategic trust in the UN system, which meant that the United States was, for the most part, a law unto itself in its administration of the Micronesian islands, and the US could use the islands for military purposes. As a result, all prying eyes were kept out of Micronesia by the US military. The citizens of countries other than the USA were completely barred from the territory and all US citizens required a security clearance from the US government in order to visit them.

The effects of a strategic trust were two-fold. Firstly, there was limited internal administration provided by the US, and secondly, the firm exclusion of outsiders from the islands allowed the unique cultures of the Micronesian groups to avoid the tidal wave of modernity that rolled over other areas of the Pacific Ocean from the Fifties into the Seventies. For the first twenty years of the US administration, the prevailing American approach was to leave the indigenous population of the islands to themselves, and consequently, no effort was made to bring the islands firmly into the twentieth century. As a result, some parts of the Micronesian islands remained a living museum of the Pacific island cultures as they had once existed, retaining much of their indigenous cultural vibrancy.

Another effect of the strategic trust was much less desirable. The strategic trust allowed the United States *carte blanche* to do what it pleased with the islands and with their people. American authority over Micronesia coincided with the Cold War between the western powers and the Soviet Union and its communist allies. The competition between the two blocs resulted in a race to develop more and more powerful nuclear weapons and it was in Micronesia that the United States began atmospheric tests of its nuclear arsenal to perfect the explosive capabilities of its bombs. The Marshall Islands, which are the chain of atolls furthest east in Micronesia, bore the brunt of the US atomic testing. During the active test period in the late Forties and the early Fifties, the United States detonated 67 nuclear bombs in the Marshall Islands. The largest test occurred on Bikini atoll, which created a fireball that was visible 400 kilometres away (and was memorialised in a type of female bathing suit designed at the same time). Children on neighbouring Micronesian islands within the fallout zone played in the radioactive ash that fell like snow on their islands. Many of the test site islands were rendered uninhabitable, perhaps for centuries to come.[1]

The isolation of the Marshall Islands, the continuing radioactivity on some of its atolls, and the reported overcrowding on Majuro, the main island, and the capital of the country, persuaded me to defer visiting the Marshall Islands during my 1989 visit to Micronesia.

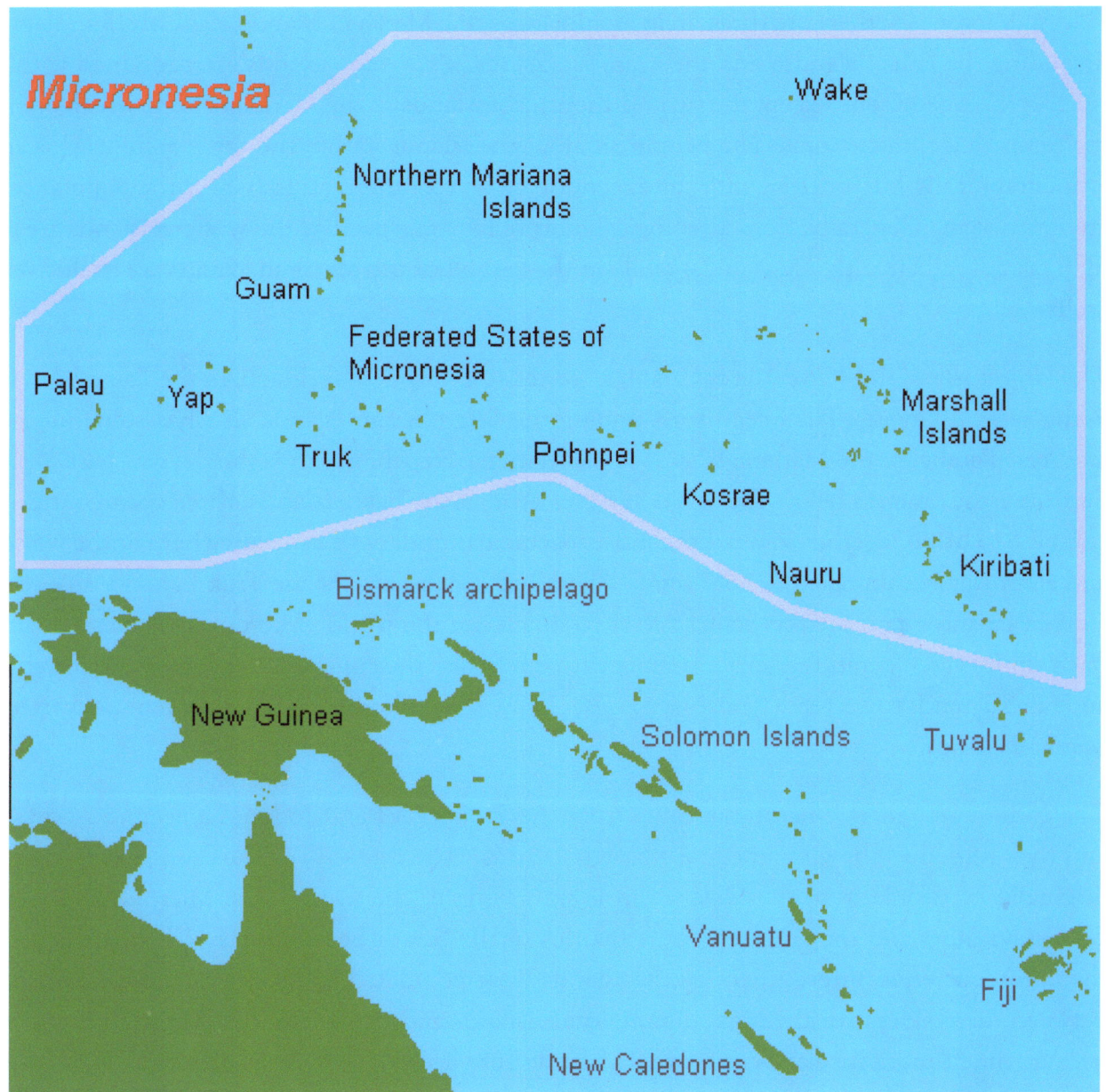

Oceanic map of the Micronesian Islands (Holger Behr/Wiki Commons public domain)

The Micronesian Islands stretch over a vast area of the North Pacific Ocean. These small specks of land are too small to be noticed by some cartographers but they extend over an ocean area that rivals the continental USA in size. The islands are warmed by blue ocean waters and have beautiful beaches, swaying palms, lush jungles of tropical plants, waterfalls, and friendly people, some of whom still live in thatched huts and in the manner of their ancestors at the time of first contact decades before. Although the total population of Micronesia around the time I visited in the late Eighties was 270,000, the population on a single island, the US territory of Guam, was 110,000, leaving just 160,000 people scattered across more than 2000 other islands in Micronesia.

Within Micronesia, there are three great archipelagos, the Marshall Islands, the Caroline Islands (including the Palau Islands), and the Mariana Islands. All of the islands are contained within the tropic climatic zone except the two northernmost islands of the Mariana chain, which are in the temperate climatic zone. The people on all of the islands within the three archipelagos are ethnographically Micronesian with the exception of two southern atolls, where the populations are Polynesian. Generally, from a geological morphology perspective, the western islands are all high islands (albeit with a few atolls) while in the east all of the Marshall Islands are exclusively atolls.

The largest island in Micronesia is Guam, a US territory, while Babeldaob, in the Palau Islands, is the second largest. The ocean surrounding the Micronesian islands also has some unique features. Parallel to the Mariana Islands is the Mariana Trench, which is the deepest part of the Earth at 10,984 metres below the ocean surface. And at the Palau Islands, three ocean currents combine to bring together some of the most spectacular underwater life anywhere in the world, and all of it is within reach of both scuba divers and snorkelers. At the Truk Lagoon, an entire sunken Japanese fleet attracts scuba divers from around the world. Micronesia is blessed with fairly uniform temperatures. From January through March the climate is at its most comfortable, and rainfall tends to decrease from east to west. However, typhoons are possible, especially in the Mariana Islands.

The earliest settlers in Micronesia came from the Philippines and Indonesia around 1500 BC, and they settled in the Caroline and Mariana islands. The Marshall Islands were settled later, originally by people from the Melanesian islands lying to the south. The Micronesians were great navigators and they crossed vast expanses of the South Seas navigating by stars and by the ocean currents. Interestingly, as the islands were populated a cultural difference emerged between the residents of the high, mountainous islands and the low atolls. The high islanders became much more insular because their islands supplied all of their basic needs. The same could not be said for the inhabitants of the small atolls and so it was the low islanders who felt compelled to travel across vast oceanic distances for trade. In so doing they helped to establish a cultural commonality across Micronesia. Each set of islands nevertheless developed its own unique customs and of the islands in general it is said that Yap Island is the most traditional of the Micronesian islands. In 1989, when I came to visit Yap, tourists were rare because the island was so far off the beaten path. Its isolation and its strong society allowed Yap to preserve much of its traditional culture unbowed by its collision with both eastern and western values and mores.

The people of Micronesia are a people of Austronesian origin and they share some of the physical characteristics with the people of the Philippines, which lie some 1200-1500 kilometres to the west. No doubt that reflects the fact that the Philippines were the source of the original wave of migration from the west into Micronesia. Like the people of the Philippines, the people of Micronesia exhibit a tanned skin tone, and dark hair, although Micronesians are, on average, taller than Filipinos, and the shape of their eyes is rounder. The Chamorro people of the Mariana Islands to the north exhibit a lighter skin tone than the Carolinian or the Palauan people, while the Micronesians of central and eastern Micronesia have some Melanesian influences in their genetic makeup probably because the original settlers who migrated into the eastern Caroline Islands from the Philippines first settled in the islands of Melanesia for several generations before they migrated north into eastern Micronesia.

A group of Micronesians on Pohnpei in 1989

Although there was a vibrant culture throughout old Micronesia the coming of Christianity caused significant cultural disruption and the knowledge of the old gods and many old legends were lost. Today there is a lack of information traceable to the oral traditions of Micronesian history. What is known about Micronesian history in the current period results from the records of the Europeans who made early contact with the peoples of Micronesia. The very first European to make contact with Micronesia was the famous Portuguese navigator Ferdinand Magellan who sailed under the Spanish flag. In 1521 Magellan arrived at Umatac Bay, on Guam, and soon afterward Spanish Galleons began to routinely stop at Guam on their transpacific journeys from South America laden with Andes silver. By the 1600s, the Spanish had sent a garrison and colonial administrators to Guam but their penetration of the remaining islands was slow and indifferent. Spanish control over nineteenth-century Micronesia remained somewhat tenuous outside of the Mariana Islands.

While Spain focussed on the Mariana Islands whalers, primarily from the United States, began visiting the eastern Caroline Islands, and they were followed by copra and bech-de-mer (sea cucumber) traders. As always when there was a collision between a less-advanced indigenous society and the representatives of western society it was difficult, or even impossible, for an indigenous society to assert its own laws and social controls in the face of western lawlessness. That was particularly true in the case of the whalers whose long periods at sea before making port at a Micronesian island primed the sailors for rowdyism. Disputes between westerners and the indigenous people were followed by local massacres of the indigenous people and that, coupled with the losses due to venereal and other communicable diseases, decimated the populations of the islands that had no natural immunity to western scourges. The populations of the Micronesian islands declined precipitously, which rendered them at risk of greater colonization. Cultural dislocation also occurred when Congregationalist missionaries from the United States established themselves in the Caroline Islands through the so-called *Morning Star* expeditions beginning in the 1850s. Earlier, Catholic Jesuits missionaries had become ensconced in the Mariana Islands and had profoundly affected the local culture in those islands. The missionaries brought western culture, and norms of morality, as well as written languages and laws but much of the past culture of the islands, which had been transmitted through oral traditions, was no longer routinely passed down to the succeeding generations.

Meanwhile, the pace of overt colonization picked up steam as the end of the nineteenth century approached. In 1878, Germany began establishing control of the Marshall Islands following an earlier penetration of those islands by German traders. The growing presence of German traders in Micronesia, particularly in the Caroline Islands, alarmed the Spanish government and in 1885, Spain and Germany agreed to appeal to the Pope for arbitration. Pope Leo XIII ruled that Spain possessed sovereignty over the Caroline Islands but that German citizens should have the right to establish plantations in the Caroline Islands and to conduct trade there. The same year Germany bought out the residual Spanish claim to the Marshall Islands and established a colony. Although Spain did not achieve its goal of barring the menacing threat of German infiltration in its Pacific Islands it did clearly establish its legal sovereignty over the Caroline Islands, and together with Spain's pre-existing control of the Mariana Islands, and the Philippines, that gave Spain a substantial Pacific empire.

Then, an event occurred that would completely undermine Spain's sovereignty over its Micronesian possessions. In 1898, the Spanish American War erupted, and the jingoistic United States of America seized both the island of Guam in the Mariana Islands and the Philippine Islands from Spain (in addition to taking Cuba and Puerto Rico from Spain in the Caribbean Sea). Realizing that, in the long run, Spain would be unlikely to hold its Micronesian possessions the Spanish government agreed to sell its sovereign rights in the Caroline and Mariana Islands to Germany in 1899, and Spain withdrew from the Pacific Ocean.

The Germans established a copra industry and began phosphorous mining in its new possessions. Germany also brought in new public health measures to reduce disease throughout Germany's Micronesian possessions. However, the German colonial era also led to the forced relocations of some island populations and to new land use policies that contributed to the outbreak of a significant rebellion on Pohnpei Island in the eastern Caroline chain in 1910, which the Germans suppressed with great severity. Then, in 1914, World War I broke out and Japan joined the war as an allied power and subsequently invaded and conquered German Micronesia with little effort. Thereafter, Japan held the islands through the war and obtained a League of Nations Mandate for the islands at the 1919 Paris Peace Conference, despite American opposition.

In the interwar years, Japan began to fortify the islands of what it called the South Seas Mandate and to integrate them into the wider Japanese Empire. All too soon settlers from Japan began to outnumber the local populations on the major islands as the Japanese undertook a massive effort to increase the agricultural production of the islands. There was considerable discriminatory treatment of the Micronesians by the Japanese but Japanese efforts to assimilate the local populations by encouraging marriages between Japanese settlers and the indigenous population were largely unsuccessful, even though a number of such marriages did in fact take place.

The island of Guam remained the only Micronesian island within the oceanic boundaries of the South Seas Mandate that was not under Japanese control. Taken by the US from Spain in 1898, Guam remained a US territory until the outbreak of World War II in the Pacific, in December, 1941. On December 10, 1941, the Japanese armed forces invaded Guam and took possession of the lightly defended island from the United States.

Three years later the US armed forces came roaring back and in 1944 they not only recovered Guam but US forces also took Kwajalein and Majuro in the Marshall Islands, Saipan and Tinian in the Mariana Islands, and Peleliu in the Palau Islands. A number of Micronesians were killed during the bombing runs and the invasions of the target islands but the United States only invaded those islands that were of strategic value to the war effort. As a result, the Japanese island garrisons on many islands were bypassed and left without resupply from Japan, and many Japanese soldiers starved. By the summer of 1945, the war in the Pacific was over after atomic bomb strikes by the US against Japan that were launched from the US air base on Tinian Island. Japan surrendered at Tokyo aboard the US battleship *Missouri* in September, 1945, and in Micronesia, the American military authorities forcibly repatriated all of the Japanese soldiers and settlers in Micronesia to Japan, even those who had entered into marriages with local women, and who had children with them.

In 1951, the administrative control of the strategic trust was transferred from the US military to the US Department of the Interior, with the exception of the Northern Mariana Islands, which remained under the control of the US Navy until 1962. The US generally followed a practice of benign neglect in Micronesia, despite its obligation to the UN Trusteeship Council to provide education, public health services, and economic development sufficient to lead the people of the mandated territory to achieve self-determination.

Eventually, however, the rapid decolonisation that overtook the European empires in Africa and Asia from the Fifties to the Seventies compelled the US to begin to move the Micronesians towards some form of local autonomy. By the mid-Sixties, the US military had begun to relax the restrictions that it had placed on visitors to the islands, and hundreds of young American Peace Corps volunteers began flooding into Micronesia. The US established high schools throughout the islands to improve the prospects of the younger generation of Micronesians. As more and more US money began to flow into the islands, however, the local population began to increasingly depend on US economic transfers.

In 1965, the Congress of Micronesia was established as the first attempt by the US to institute a form of self-government within the Trust Territory. To further a sense of national belonging Peace Corps volunteers were sent to every inhabited island in order to foster the use of English as a national language for Micronesia. However, quite contrary to the expectation of the United States that Micronesia would form a single polity, the indigenous peoples determined that the various island chains would go their separate ways. Consequently, the Northern Mariana Islands opted to link its future directly to the USA as a self-governing US commonwealth, similar to Puerto Rico, while the Caroline, Marshall, and Palau islands all separately moved towards formal independence, though retaining strong economic and military links to the USA.

Since the presidency of Theodore Roosevelt the United States had cause to rue the day that it did not take possession of all of Micronesia following the Spanish-American War and now, as the islands moved closer to autonomy, the US determined to keep its economic and military leverage over the islands through the US economic support that the islanders had come to expect. The United States achieved that objective by negotiating and signing Compacts of Free Association in 1982, with the Marshall, Caroline, and Palau islands. Those compacts tied the islands to the USA economically and militarily, by allowing the US to guide the foreign policies of the new island states, and ensured the continuing military domination of Micronesia by the US military.

As the move towards self-government progressed through the Eighties the Micronesians were increasingly finding themselves caught between their traditional cultures and modernity.

NOTES

[1] Out of a sense of guilt perhaps, or to ensure that close relations between the Marshall Islands and the United States would continue, the US agreed in the Compact of Free Association that it signed with the new Republic of the Marshall Islands in 1986 that the Marshallese could live and work in the United States and be exempt from immigration control. Initially, the United States also paid for health care for Marshallese living in the United States but it later revoked that privilege, leaving the Marshallese as the poorest ethnic community in the United States, according to US census data.

6

YAP – THE ISLAND OF TRADITION

On February 4, 1989, I flew from Saskatoon, Saskatchewan to San Francisco, California via Vancouver, British Columbia on Canadian Airlines and stayed overnight at the Radisson San Francisco Airport Hotel. The next day I found myself in the large arc-shaped departure terminal at the San Francisco International Airport awaiting a Continental Airlines flight across the Pacific Ocean to Micronesia. Soon I heard the airline staff over the intercom system welcoming passengers to "Continental Airlines' Transpacific Service, with service to Honolulu Hawaii, and onward service to Japan". I boarded Continental Flight No. 41, a Continental DC-10, which departed on February 5th on a 5-hour flight to the Hawaiian Islands.

At Honolulu, I transferred to Continental Flight No. 3, which departed later that day for a seven-hour flight to Guam, the transportation hub of Micronesia. The flight to Guam was very lively, particularly because there was a pub in the forward upper deck of the aeroplane giving passengers a chance to mingle with fellow travellers. I met a group of divers from Winnipeg, Manitoba, who were going out to Micronesia for what they hoped would be the diving experience of a lifetime. I also met a couple of men from the Battelle Consulting Engineering firm in Denver, Colorado who were going out to Micronesia on a U.S. Government contract. They explained to me that in the run-up to the end of the Trust Territory administration the US Government had begun a large-scale capital investment program in Micronesia to help the new island nations ready themselves to emerge onto the world stage (and, as well, to continue to tie the newly independent countries closer to the United States). Since we were chasing the Sun across the Pacific it was not yet dark when we arrived at Guam. But I had now crossed the International Date Line and I had lost a day from the calendar, so I arrived in Guam on the evening of February 6, 1989.

The busy aeroport on Guam is the air transportation hub of Micronesia

I awoke early on the morning of February 7, 1989, eager to begin my Micronesian adventure by journeying to the island of Yap. Carol, my travel agent back in Saskatoon, Saskatchewan had booked me an independent tour through a tour company that specialized in Micronesia called Trip-n-tour and she had made sure of all my flight arrangements between the islands.

At the aeroport, in Guam, I checked in with Continental Air Micronesia the local inter-island air carrier that was universally known to the locals as Air Mike. This was my first experience flying with the local carrier in which Continental Airlines held a 30% stake. Air Mike flew the older model Boeing 727s, such as B-727-24C and the B-727-92C, which were no longer used in passenger services in North America. The planes had open luggage racks above the seats, which clearly marked them as older models, and they also had embarkation ramps at the rear of the aeroplane that were frequently used to embark or disembark passengers at the local airports in Micronesia. The use of the rear ramp was necessary because Air Mike's aeroplanes were configured to carry both cargo and passengers, and the cargo hold was built into the forward section of the aeroplane leaving the rear ramp as the main entry and exit portal on the aeroplane.

After settling into my seat I leafed through a copy of Air Mike's in-flight magazine which was called appropriately 'Islands.' The edition I had (vol. 8 no. 1, of 1989) was 56 pages in length of which only seven and one-half pages were written in English while the remainder was in Japanese. That certainly told me where Air Mike's bread and butter tourist traffic was originating from. Only a few tourists from North America found their way into Micronesia while Japanese tourists were common, especially on Guam and Saipan. Yap, however, attracted very few tourists from anywhere. Although Yap was known to be a particularly good place to scuba dive because of the large seagrass meadows offshore, the main attraction offered by Yap was its unique and still vibrant Pacific island culture.

Yap is 829 kilometres southwest of Guam and together with its outlying islands forms the westernmost state of the Federated States of Micronesia ("FSM") which incorporates almost all of the Caroline Islands. Yap actually consists of four separate islands of which the island of Yap proper, which is called Waab in the local Yapese language, is the main island in the group. The other three islands of Yap are called Tamil-Gagil, Map, and Rumung, and there are also ten smaller islands within the reef that surround Yap. Some 65% of the population lives on Yap proper, the main island, with the remainder of the population living on the other three islands.

The flight from Guam to Yap was uneventful. Once I landed in Yap I discovered that the Yap Aeroport Terminal was just an old Quonset hut with a tin roof that must have been left over from World War Two. Inside, the structure was filled with unpainted wooden structures that had been weathered grey, like barn-board. Woven dividers extending part way up provided a partial division between different stations. The local customs and immigration officer for the Federated States of Micronesia who was on duty when I arrived was a fat shirtless guy sitting behind an unpainted and weathered grey wooden frame station that was enclosed by chicken wire. He was also chewing a betel nut that had turned both his saliva and his teeth red. He was not at all friendly. He looked at my passport and then rifled through my luggage, which I had perched on a rickety wooden bench, also weathered and grey. Beneath me, the floor was just dirt covered with red splotches, the residue of expectorated betel nut saliva. Finding nothing remarkable about me, or my luggage, the officer merely grunted and stamped my passport with an FSM entry visa. As I left the airport I felt like I had entered a time warp and that I had perhaps landed in the 1930s, but I soon noticed that next to the beat-up Quonset hut terminal a new modern air terminal was under construction with a large sign announcing on behalf of US President Ronald Reagan that the United States was building a new aeroport terminal on Yap for the FSM.

At the aeroport, I was picked up by a van operated by the ESA Hotel, one of only two small hotels operating on the island in those days. The two hotels on Yap in 1989 possessed a grand total of 26 rooms between them so Yap was clearly off the beaten path for tourism. My tour arrangements placed me at the ESA Hotel, which was run by a Palauan family that had settled on Yap. The ESA looked out over Chamorro Bay in Colonia, the capital town of Yap and it possessed sixteen clean rooms with private bathrooms, air conditioning, and television. There was also an open-air central lounge with a television for the use of guests.

The rooms at the ESA were comfortable but spartan. A small gecko lived inside my room and he spent much of his time running up and down the curtains on my window. Outside the room was a common balcony where guests could sit on plastic chairs overlooking the water. A restaurant was located inside the hotel downstairs, offering adequate and reasonably priced American-style meals. Cable television had recently come to Yap and guests of the ESA Hotel were able to watch American television. However, the American television that was available on Yap was sent out to Micronesia from the United States on videotapes, and it took upwards of three weeks for the videotapes to reach Micronesia before they could be played on the local island cable network. Therefore, it was three-week-old television that was available on Yap. When I turned on the television in my room I found myself watching the US television coverage of the inauguration of President George H.W. Bush, an event that had actually occurred three weeks before I arrived on Yap. I thought to myself here was a place that had certainly not quite caught up to the rest of the twentieth century. But the influences of the modern world were coming fast and I realized I had come at the right time before modernity brought even more significant changes to this island.

The view from the common balcony at the ESA Hotel on Yap Island, 1989

At the hotel, I met two American scuba divers who were travelling together and had come to Yap to explore the reef surrounding the islands. Jim was in his early sixties and Susan, the woman who was his travelling companion, appeared to be in her mid-forties. He was the senior scientist for a major US chemical manufacturer, and she was a former nurse who was now managing a marina on the US mainland. They both enjoyed scuba diving in the more unspoiled areas of the world and they often travelled together to visit some of the world's most remote diving spots. Because Jim's function was of critical importance to the company he worked for he told me that each time he went on a diving expedition he was required by his employer to meet with the company's medical advisor to review a list of health dangers and possible diseases that he might encounter, as well as the precautions that he must take. Micronesia was not too bad from a health risk perspective, according to Jim, but he said that when he visited Papua New Guinea the list of possible health hazards that he faced was so long that the company doctor just threw up his hands and told him "Whatever you do, don't drink the water!" I thought that might be good advice for Micronesia as well, and I made a mental note to avoid the local tap water.

Generally, there were no major health concerns about traveling in Micronesia provided that one took some basic precautions. The water was not safe in many places due to giardia and amoebic dysentery, so I planned on drinking only canned beverages, bottled water or drinking coconuts (and beer). Although I learned that Truk Lagoon had experienced a cholera outbreak in 1983-84, that outbreak was over long before I travelled to Micronesia and I was not planning to spend much time on Truk so that was not a particular concern. Dengue Fever was present but rare in Micronesia, and some of the Polynesians currently living in Pohnpei who had been displaced from atolls further south by cyclones had been exposed to Hansen's disease, otherwise known as leprosy. But the most likely health danger I would probably encounter was sunstroke, which meant that I had to always ensure that I had access to liquids. All in all, I thought that any health issues that I would encounter were acceptable risks, provided that I was careful.

Health issues are always a concern when one travels to remote areas and on Yap only limited medical services were available. And what was available on Yap was not the same as one might expect back home in North America. That was clearly brought home to me one afternoon while I was taking in a leisurely view from the common balcony of the ESA hotel. Suddenly, the door of the room next to mine banged open and a US Coast Guard officer emerged, sweating profusely and barely able to stand. He begged me to get some medical assistance for him immediately. Fortunately, my diving acquaintances were in the hotel that afternoon and I remembered that Susan was a former nurse so I sought her help for the Coast Guard man. Susan did a quick assessment and then sent for a US Navy Corpsman from the US Navy Seabee base on the outskirts of Colonia. Soon after I found Susan and the Corpsman with the unfortunate Coast Guard officer in the common room of the hotel where they had hooked him up to portable medical equipment used for pumping stomachs, which they then proceeded to do right in the hotel's common area. Later, I was told that the man had been diagnosed with dysentery, a disease that occurs with some frequency in Micronesia, and he had to be medically evacuated to Guam for further treatment.

Among the various other characters staying at the hotel was a heavyset older Caucasian man who was a commercial trader within the islands. Dressed in short pants and a summer blazer, with a broad-brimmed hat perched on his head he made me think instantly of the South Seas traders who had appeared in some of the literature that I had read. His role in the islands was to arrange to supply stores and other commercial establishments with the products they needed to sell to the island residents throughout Micronesia. Bulk supplies came into Yap either by a chartered freighter or by the government-owned *MV Micro Spirit* – a Federated States of Micronesia owned-ship that provided inter-island transport services between the various island groups in the country. In Colonia, the main emporium was the Yap Co-op store, a sort of South Seas general store, and it was the place where it seemed everybody went shopping and exchanged news or gossip.

The South Seas trader was not the only trader circulating on Yap while I was there. I also met a Filipino man, Rodolfo Juarez, who worked for Pepsi Micronesia, and who was engaged to supply Pepsi soft drinks throughout the islands. When I visited the Federated States of Micronesia in 1989, the country may have been unique in the western world because one could not buy a bottle or can of Coca-Cola anywhere. Pepsi-Cola had an absolute lock on the local soft-drink market. If you wanted a soft drink you had to buy a Pepsi. The Filipino fellow I met routinely travelled throughout Micronesia to maintain Pepsi's dominance in the market. I asked Rodolfo what had drawn him from the Philippines to Micronesia and he explained to me that back in the Philippines he had a job with the police serving under the dictatorial Marcos regime but his superiors required him to do certain acts that had violated his conscience, and so he had decided to leave the police and to leave the Philippines. Those were the days when the dying regime of the US-backed Philippine dictator Ferdinand Marcos perpetuated human rights abuses on Philippine dissidents. Eventually, Marcos' abusive tenure proved too much for the people of the Philippines and he was ousted from power. As for my acquaintance, I could see that even the memory of his police experiences raised difficult emotions. I assured Rodolfo that from my perspective I thought he had done the right thing to leave such a career behind him. He gave me a blue Pepsi Micronesia T-shirt as a memento of our meeting.

Pepsi was not the only beverage with a monopolistic lock on the local Micronesian market. I also found that the only beer that Micronesians on Yap would drink was Budweiser beer, and the evidence of that was all over the island in the form of discarded Budweiser beer cans. Apparently, the people of Micronesia, and the Yapese in particular, preferred Budweiser because Budweiser was touted as 'the King of Beers' and why would one drink anything less than the king of beers?

One morning I hiked up Medeqdeq Hill outside Colonia, which brought me to a height of 450 metres above sea level. I was able to gain a good view of the capital town, as well as a large part of Yap Island itself. Unlike many other South Seas islands, Yap was not formed through volcanic action but rather by an uplifting of a continental shelf. It has the lushness expected of the South Seas, however, with a consistent temperature of 27 degrees Celsius, and a 12-degree Celsius temperature variation from night until noon. There was significant humidity at night as well as in the morning, although rains tended to come later in the day. February, when I visited the island, is the driest month of the year and I found the temperature to be quite pleasant.

Overlooking Colonia, Yap, from Medeqdeq Hill, 1989

The people of Yap are of Micronesian stock, Austronesian, with some Philippine and Indonesian blood. Before the coming of Christianity, the Yapese dominated the neighbouring islands because of the purported ability of Yap sorcerers to cause famine and sickness. In modern times Yap became world-famous for its stone money, which consists of large round stones that were, and still are, used as a form of currency. Yap's stone money constitutes the largest (and heaviest) currency in the world and most of it is kept in so-called stone money banks, which consist of block-long collections of stone money that are half buried in the ground lying parallel to certain roads near Colonia, and in other parts of the island, particularly in an area that is called Rull.

A stone money bank on Yap, 1989

The first western contact with Yap came in 1526 when a Portuguese explorer named de Rocha arrived. He found the natives to be friendly and he named the island Yap, mistakenly thinking that was the name of the island when in fact the word *yap* was the local word for a paddle![1] In later years the Spanish influence deepened through trade between the people of Yap with the Mariana Islands but the most frequent nineteenth-century contact between the Yapese and westerners was through contact with British traders. However, the natives of Yap were considered to be unfriendly throughout much of the nineteenth century and some westerners were actually massacred on Yap so it was not until the 1860's that regular trade with the island was established. In 1860, a German trading compound was established, and in 1871 an Irish-American named David O'Keefe established a copra trade on Yap by transporting quarried stone for the manufacture of stone money from Palau to Yap in return for local copra, a trade deal that helped to make him a rich man. Spain finally asserted its previously declared colonial rights in 1885 and established a garrison on Yap that was prompted by Spain's fear that Germany was undermining the tenuous Spanish colonial presence in Micronesia through trade. Conflicts over the German economic penetration of Micronesia led to a papal arbitration of the relative Spanish and German territorial claims in 1886, in which the Spanish were successful. Spanish colonial administration lingered thereafter with varying degrees of effectiveness until the Spanish-American War in 1898, which saw Spain lose the most important part of its Pacific empire, the Philippine Islands, to the United States. With little to be gained by holding onto the Caroline Islands in the aftermath of that war, the Spanish agreed to sell the islands to Germany in 1899.

After taking control Germany made substantial investments in Yap by building the Tagreng Canal to provide inland boat travel between Colonia and the northern part of Yap Island, as well as establishing a long-range wireless transmission station in 1905, to connect Berlin with its overseas Pacific empire. The Germans also brought with them modern ideas and techniques of public health, which improved the living conditions of the inhabitants of Yap. However, the Germans also compelled forced labour from the Yapese. The German colonial period only lasted fifteen years, until the outbreak of World War One. In 1914, shortly after the war began Japan joined the allied powers and sent its navy to Micronesia to oust the Germans, and take possession of the islands for Japan. At the Paris Peace Conference in 1919, Japan's control of Micronesia was confirmed when Japan was awarded a League of Nations mandate over the islands. The United States was unhappy about the Japanese acquiring a mandate in Micronesia fearing correctly that it would give Japan the ability to cut US communications with America's Philippine Islands colony. The US was particularly concerned about the Japanese control of Yap, which continued to be an important relay station for trans-Pacific cables in the post-war world. To ameliorate the US concerns Japan signed a treaty with the United States in 1921, which allowed some American access to Yap to maintain the important USA-Shanghai trans-Pacific cable.

Within its mandated Micronesian territories, Japan displayed no intention of preparing the native people for the ultimate stage of independence. Rather, Japan intended to fully integrate Micronesia into the Japanese Empire and, for that purpose, Japan sent settlers to Yap to establish commercial enterprises, and to promote agriculture. Soon Japanese settlers outnumbered the natives although efforts by the Japanese to promote the intermarriage of the indigenous population with the Japanese, and thus further assimilation, came to nought. In the final interwar years, the Japanese forced the Yapese to build airfields and fortifications and the failures on the part of the locals to meet Japanese demands led to the Japanese smashing the traditional Yapese stone money and throwing the remnants into local road fill. During the war, Yap was bombed by the Americans but it was not invaded. US troops only arrived in 1945, after the end of the war. The United Nations included Yap in the strategic trust that it granted to the United States after the war and Yap and the other Micronesian islands disappeared behind a tightly controlled US military administration that excluded most outsiders from the islands. Later the islands fell under the administration of the US Department of the Interior but even then tourism remained discouraged. Although the indigenous population of Yap declined significantly under Japanese rule it subsequently recovered under the US administration and by 1980 there were approximately 10,000 people in Yap.

The main centre on the island, and the state capital of Yap state, is the town of Colonia, which was named by the Germans when they ruled here. Colonia is an amalgam of motorcycles, pick-up trucks, and cable television with grass skirts and loincloths thrown into the mix. Much of its society in 1989 remained fixed in the past. Yap exhibited a relaxed and simple island lifestyle with its customs, architecture, and manners of personal dress harkening back to pre-modern times. Here I found a South Seas island culture that had not been greatly disrupted by its cultural collision with modernity. The tide of modernization was washing over Micronesia but Yap had not yet succumbed and so my visit to Yap was a fortunate introduction to a unique island culture.

The dress of the men of Colonia for the most part consisted of western-style pants, either long or short and a western-style shirt. However, there were still to be seen in Colonia men who wore the traditional *Thu*, a loincloth that was usually blue but sometimes red in colour, and who were otherwise naked. The *Thu* was made of cotton cloth and was stretched tight between a man's legs and then tied around the waistline. The loose ends were then draped in front and in the back and left swinging. Although the *Thu* was no longer typical by 1987, particularly among the residents of Colonia, it could still be seen on male visitors arriving in the town from the outer islands.

There were strict customs regulating female modesty in Yap but the modesty customs of Yapese culture diverged considerably from western culture. The appearance in public of bare-breasted women was simply seen as normal and acceptable throughout Yap, although there was a strong cultural aversion to a woman exposing her thighs. Thus, women on Yap, including Colonia, could be seen walking abroad in public with uncovered breasts but uncovered thighs were forbidden. Female swimsuits that revealed the thighs were acceptable only while a woman was actually in the water, and her thighs had to be covered as soon as a woman emerged onto the shore. Those who violated this norm (invariably a western visitor) were, at a minimum, often subjected to rude comments.

Thus, while western-style fashions were making inroads in 1989, it was still quite common to see bare-breasted women going about their business in town. I saw bare-breasted women shopping in the Yap Co-op store, and even performing such mundane tasks as pushing a lawnmower to cut a lawn at a residence. To ensure that their thighs were modestly covered, however, most women wore a skirt called a lava-lava. Traditionally, the lava-lavas had been made from the shredded inner bark of the hibiscus plant, or from banana tree fibre, but they were now made from cotton, which Yapese women found to be easier to work with when making them. The cotton lava-lavas were also much more comfortable to wear because they were less scratchy than the lava-lavas made from hibiscus bark. A lava-lava was worn by wrapping it around the lower body of a woman and tying it in place. However, women who worked in commercial establishments, or public offices, generally dressed western-style. Grass skirts made from the hibiscus plant were reserved only for the participation of women in traditional cultural ceremonies.

One day, I was visiting the Yapese Women's Association store ("YWA") shopping for handicrafts and I spoke to the young female clerk (who was dressed western-style) about the topless female custom on Yap. She explained to me that on Yap the women find it quite normal to expose their breasts in public and they encountered no issues about it from the males in their society. She said, much to my surprise, that some Yapese women even went topless in the local Roman

Catholic Church on their wedding day! However, she emphasized that on Yap women must cover their thighs because it violated taboo when women displayed their upper thighs in public. Susan, my female American diver acquaintance, told me that when she stripped off her wetsuit at the dock following a dive and revealed herself in a one-piece bathing suit, she immediately began to hear mutterings about her naked thighs from the older Yapese women who were present on the dock, causing Susan to quickly cover up her legs. Another fashion accessory often seen at cultural ceremonies was the Yapese *maramars*, a type of floral coronet worn to signify happiness and love. I saw it worn as part of the Yapese ceremonial dress.

In addition to their sartorial choices, there was one other remarkable cultural custom among the Yapese. Everybody chewed the betel nut, which is the nut of the Areca Palm, which the locals called *buw*. The Yapese chew betel nut constantly because it is a mild narcotic but the mild high achieved by chewing the nut only lasts about 10 minutes as I found out for myself when I tried it. The preparation of this addictive delicacy involves splitting the betel nut open while it is green and sprinkling dry lime derived from local coral over the contents before wrapping it in a pepper leaf. It is then ready to be chewed. One result of this practice is to produce an excess of bright red saliva in the mouths of those chewing the nut, which then turns the chewer's teeth a bright red colour. As a variation, some people mixed tobacco into the preparation and that turned their saliva dark, leaving them with black stains on their teeth. The excess saliva produced by chewing *buw* was generally expectorated anywhere nearby and thus the heavily travelled roads in Colonia everywhere bore red splotches wherever one looked. It seemed impossible to overestimate the use of this narcotic, and the vast majority of the local population partook of it.

Villages on Yap were organized according to a caste system with the members of the same caste inhabiting the same village. There are nine castes on Yap and each village has a chief, and above a village chief are the three paramount chiefs drawn from the highest-status castes for the entire island group. A caste does not necessarily determine a person's standard of living but it does denote their civil status within the hierarchical caste structure of the island community. The caste system reinforced marriages within the Micronesian community and was one reason why there were few mixed Micronesian-Japanese marriages during the Japanese colonial period, despite the persistent efforts of the Japanese to actively promote such unions. One result of the caste system was that people of the higher castes expected free labour from those who were members of the lower castes. There was also a universal rule applicable to all castes that women held a lesser status than men. Despite the lesser status of women, however, all Micronesian society was organized on a matrilineal basis.

The existence of castes on Yap did not affect the democratic principle that was applied to the elections of state and national leaders in the Federated States of Micronesia, and there was universal adult suffrage. However, the traditional chiefs did have a significant influence on the elections. Furthermore, there were two councils of traditional leaders on Yap and those councils were empowered to veto laws affecting the traditions and customs of Yap, so Yap might be called a limited democracy at best.

I had been forewarned about visitor etiquette on Yap. The people of Yap are sensitive about having their photographs taken because the Yapese believe that the act of photography captures one's spirit. Therefore, it was important that visitors ask for permission before taking photographs of a person (unless taking a photograph at a public ceremonial event). Property rights are also sacrosanct on Yap and every piece of land is both marked and named with the village chiefs owning the highest-rated plots of land. Therefore, I was warned that when walking about visitors should not enter a village, or trespass onto private property, unless the village chief or the owner of the land has given their permission to enter the village or the property. In some circumstances, such as when the chief, or landowner, is away it was acceptable to ask the nearest adult for permission to intrude. I was even told that I should ask for adult permission, if possible, before taking photographs of private property because of the importance of property rights to the Yapese. There were, however, no restrictions on the use or photography of the public roads, which belonged to all, nor were there any restrictions on entering a commercial establishment or public office.

The landscape of Yap Island, 1989

Overall, I found the Yapese to be a proud but shy people living in traditional communities that sought to integrate the desirable aspects of modernity without being overwhelmed by it – a rare place where respectful visitors could experience a uniquely preserved oceanic island people and their culture.

Outside of Colonia the terrain of the interior of Yap consisted of rolling hills covered with grasses interspersed with occasional palm and pandanus trees. Along the coast, the foliage was denser and lusher, and there were quiet villages that almost seemed deserted during the day. My recollection of the countryside out of Colonia is of peaceful villages with old stone pathways lined with palm trees and hibiscus plants. Stone walls that were not built too high marked the all-important property lines in a society where property rights were so very important. Within the villages, the houses were made of wood, thatch, rope, and bamboo.

Much of the economy was based upon subsistence agriculture, and in such a bountiful environment it was possible to grow taro, yams, limes, sweet potatoes, watermelon, bananas, cocoa, Polynesian chestnuts, breadfruit, oranges, tapioca, papaya and, of course, coconuts. The people were also excellent fishers, using line and hooks, fishnets, spears, and fish traps. They sailed in outrigger paddling canoes, and outrigger sailing canoes, which they kept protected in canoe houses.

Within the small market economy of Yap, the main export was copra. There were also exports of trochus, a type of sea snail delicacy. Tourism and handicrafts formed a very small part of the local economy. Where salaried positions existed about 75% of those positions were dependent upon funding from the US Government, which showed how tightly integrated economically Yap was with the United States.

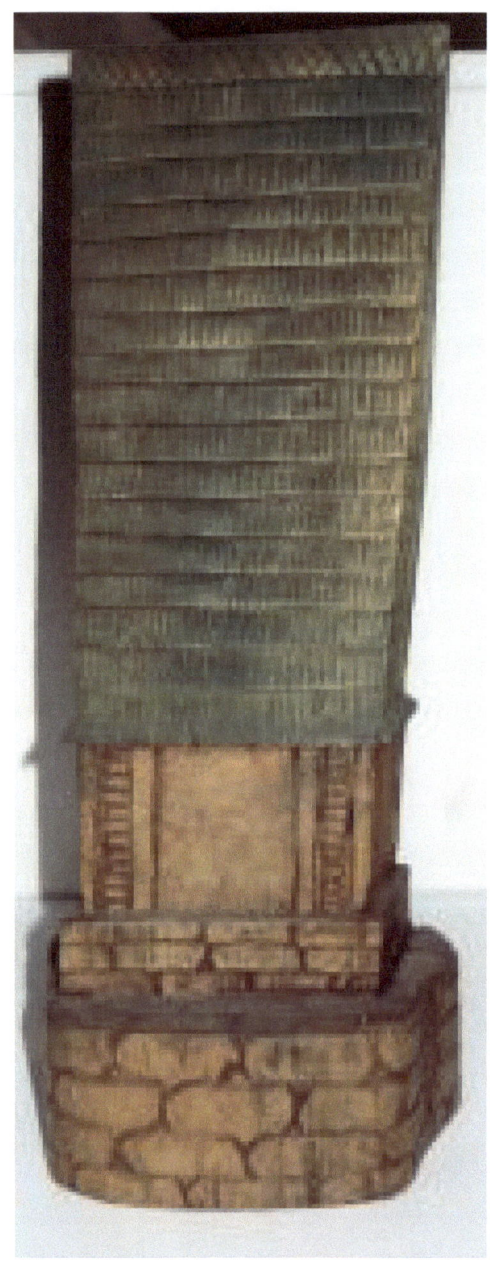

A *Yap carving of a Men's House. Handicrafts were a small part of the economy in 1989*

A Storyboard from Yap, depicting a battle scene

Details from the Yapese Storyboard

While on the surface there was a degree of modernity underneath Yap remained very traditional. And yet, although Yap was a place that had avoided or resisted wholesale modernization change was in the offing. Cable TV had recently come to Yap and noticeable social changes could not be far behind, particularly in the matter of dress where western-style clothing was becoming more commonplace in preference to traditional attire. Still, the Yapese remained the most traditional of all Micronesians, and this island indeed remained the heart of traditional Micronesian culture.

Yap did not have many beaches owing to the extensive growth of mangroves along the coastline and perhaps that was a blessing because it prevented rampant tourism development from overrunning the island. I took a motor tour of the island with the manager of the ESA hotel, a Palauan whose family had settled on Yap. One of the first places we explored was the stone money banks close to Colonia. A short distance out of Colonia there was a good example of this unique currency of Yap. Large and small stone wheels were embedded in the soil along the side of the road. Historically, the stone money of Yap was quarried from calcite deposits in the Palau Islands. A long time ago a legendary navigator found stones of the right type to render into flat wheels with a central doughnut-type hole in Palau. Yapese mariners continued to sail to the Palau Islands, some 400 kilometres from Yap, to quarry the stone and bring it back to Yap where it was fashioned into the giant circular stone wheels that served as the traditional currency on Yap. The journey and the effort to obtain the stone was very dangerous owing to the great distance of ocean between Yap and the Palau Islands, and it was made more dangerous by the weight of the rock, the possibility of storms, and the limited size and design of the sailing canoes that transported the rock to Yap. Many sailors were lost at sea while bringing the desired rock to Yap and, as a result, the most valuable stones are the ones that cost the most lives to procure. Sometimes the names of the mariners who lost their lives in bringing the stone back to Yap are inscribed on the carved stone money.

The Yapese call their stone money *rai* and although most of it is of a fairly modest size some *rai* can be more than three and one-half metres in diameter, and weigh several tonnes. Nowadays, traditional stone money tends to be fixed in one place in a stone money 'bank,' where the wheels of stone are embedded into the ground. Although the ownership of traditional stone money can change, the actual physical location of stone money embedded in a stone money bank does not usually change. Although stone money is no longer coined there is still plenty of it on the island. In addition to the stone money banks found alongside some roads, stone currency can be found on private property and also around Men's Houses, or Community Meeting Houses, across Yap. During the Japanese colonial period, there were a total of 13,281 stone coins according to a 1929 census of the stone money conducted by the Japanese colonial administration. However, some of the coins were subsequently lost (or destroyed by the Japanese) so it was uncertain how much stone money actually remained when I visited Yap in 1989. According to my Palauan guide, stone money was still being used in 1989 to settle traditional debts but the US greenback was now the universal legal tender throughout all of the Federated States of Micronesia, including Yap.

We drove south of Colonia over rough dirt tracks past the high school to the old German wireless station. In 1905, before World War One Germany built a 90-metre radio tower on Yap, which formed part of Imperial Germany's world-wide communications network linking its overseas colonies in China, the Pacific Ocean, and Africa, with Berlin. When the world war broke out in August, 1914, the British Royal Navy sent the cruisers HMS *Minotaur* and HMS *Newcastle* to Yap to bombard the wireless station and disrupt German communications. The two cruisers shelled the radio station and put it out of action on August 12, 1914. When I visited the site in 1989 the ruins of the station were still standing. Although the Germans made some efforts to repair the wireless station after the British attack the station was put out of action permanently when the Japanese Navy subsequently occupied Yap later in 1914. We drove on further up into some fields near the airport where I encountered the remains of the next world war in the form of the wreckage of a Japanese Zero warplane that had been strafed and destroyed by US aviators during World War II when Yap was an important Japanese base in Micronesia.

A destroyed Japanese Zero fighter from World War Two on Yap

The coastal road on the west side of the island was a mangrove coast, and there were fruit bats roosting in the trees. We stopped at a village where there was a *faluw*, or Men's House. After getting permission from the chief to visit we followed a stone path to visit the *faluw*. A raised stone platform supported pole structures which held up a palm-thatched roof somewhat in the design of an A-frame building, with the front of the thatched building pitched forward. The roof was constructed with the leaves of pandanus, nipa, and coconut, and bamboo was used to support the roof. The floor was wooden, constructed of treated betel nut wood and, surprisingly, the structure was wired for electricity. Coconut fibre was used to tie some of the architectural elements together. Outside, several examples of stone money leaned against the foot of the platform. Here in the *faluw,* the village men would traditionally meet to discuss affairs relating to village life and it was also here where young men and boys would be mentored for their future roles as providers for their families, and defenders of their village. Rites of passage for males were practised here, and male social linkages were established and maintained. Young men would be trained to fight, fish, build canoes, and navigate as well as learn their proper social roles and conduct within Yapese traditions, customs, and music. Women were expressly excluded from entry to all male meetings in the *faluw*.

A view of a faluw or Men's House on Yap Island, 1989

However, as a consequence of the formal education that was now being provided by the American-supported governmental institutions, the educational function of the *faluw* had declined, and a more club-like purpose of the meeting place was becoming increasingly dominant. Meetings now tended to focus more on village politics, fishing and community affairs rather than the mentoring and rites of passage of the young.

A faluw or Men's House on Yap Island erected on the shoreline of Yap

The location of a *faluw* is always chosen to be a site that is close to the shoreline in order to provide access to the ocean for fishing and for visitors to arrive by outrigger canoe without trespassing on private property. When planning the construction of a *faluw* the heads of high-caste households would meet to determine what materials were required for the construction of a *faluw* and which high-caste member should be responsible for obtaining the requisite materials. Then each high caste man would pass down to the lower caste men beholden to them their particular requirements for constructing the *faluw* and those men would go about providing them. A construction advisor with experience in *faluw* construction would then be appointed and the construction work would begin.

An interior view of a faluw. Note the presence of electrical lighting, a modern convenience.

There was also another type of male meeting place called the *wunbey*, which was an open-air setting with a central raised platform and stone backrests for sitting elders. The male elders would sit against the stone backrests while a low platform in the centre held food and betel nuts. The younger men, sitting on the edges of the platform, would serve those delicacies to elders.

The open-air meeting place, or wumbey, used by male village elders on Yap Island

For meetings of an entire village community a thatched communal meeting house called a *pebai* was built with open sides, thus distinguishing it from the Men's House. In this structure, all members of the community, both male and female, could be present. Finally, there was a special Women's House that was a place of sanctuary for women who were experiencing their menstrual cycle. Traditionally, the women would go to the Women's house for a few quiet days while they were menstruating, and then return to their homes in the village. I was told, however, that with increasing modernity this custom had greatly receded on Yap, although it was still very much adhered to on the outer atolls of Yap State. All of these special houses that I viewed were built adjacent to the ocean, an invariable custom of the Yapese.

A structure on Yap that was described to me as a Women's House once used by females during their menstrual period. This custom had disappeared on Yap by 1989 and the structure was now repurposed for community meetings.

My guide explained that traditional ways are still very prevalent in Yap and it has retained its strong culture, with the traditional chiefs still playing a strong role in Yapese society. As a result, there was no need for state welfare on Yap because families would take care of their own, always. He compared Yap favourably to some other oceanic places, such as the Philippines, where serious social problems had recently emerged, including some media reports of underage girls being sold into prostitution.

While we drove along the coast I asked my friend what he liked to do for recreation and he told me that he enjoyed spearfishing, particularly at night. In fact, he said that he had been out in the ocean doing just that the night before our trip around the island. He explained that he and his friends typically go out into an area of the ocean off of Yap where they know fish tend to congregate and then dive into the water to hunt their prey with spears. I knew that sharks tended to feed more often at night than in the daytime and so I asked him if he was worried about the danger of sharks. Wasn't he afraid of them? He told me that in his experience the sharks will not bother people even when people are spearfishing but if the sharks become too curious his practice is to shine a flashlight directly into a shark's eye and immediately shine the flashlight down into the depths. When that is done, he said, invariably the shark will follow the light down into the depths, leaving the diver to engage in spearfishing unmolested.

Beaches are limited on Yap owing to the extensive mangrove vegetation along the coastline

Back in Colonia, I stopped by the local office of the US Peace Corps Agency and talked to some of the young American Peace Corps volunteers about their life on Yap, and the efforts they were making to assist the indigenous people. While I was there I also happened to meet two young men from the US Navy Seabee Base on Yap. I struck up a conversation with them and it turned out that one of them had married a Newfoundland girl. They were in a happy mood because their team was preparing to ship out of Yap on rotation. A feast was being planned by the Yapese to mark the team's rotation out of Yap and my new acquaintances invited me to attend their farewell feast at the Seabee base later that day. I was happy to accept their invitation, particularly because they told me that there would be performances of the colourful Yapese stick dance at the farewell feast. And so, later that afternoon, I made my way out to the Seabee Base just past the local high school in Colonia.

The Seabees are a US Navy Construction Battalion whose personnel provide construction and technical services to the Navy and Marine Corps and one Seabee unit was maintained on Yap Island. I discovered that the Seabee base was a collection of small bungalows with louvred windows and metal storage huts, replete with a communications tower, a traditional naval flag mast, and a small grass lawn area. All roads leading to it were gravelled but the roads inside the small base area were just dirt tracks.

Outgoing and incoming Seabee detachments muster at the Seabee Base on Yap Island, 1989

The changeover ceremony was attended by local Micronesian dignitaries and American officials, as well as some invited guests, such as myself. The formal ceremony began with the departing Seabees mustered on parade and draped with flower leis as short speeches were given by the local civic officials. That was followed by a formal response on behalf of the departing Seabees. The replacement team of Seabees was also mustered for the ceremony. Following the speeches, two Yapese dance groups consisting of local youths moved slowly into the base area, and each of them, in turn, performed a traditional Yapese stick dance accompanied by chants and shouts. The stick dance or *churu* is a dance of ancient provenance and is intended to commemorate long ago battles during tribal warfare. Essentially, the stick dance is a form of ancient military training. On Yap, it is performed by both sexes but never in mixed groups unless the participants are children or youths. In the dance performances that I witnessed there was one performance by a mixed group of youths and another one by a group of young women.

The first group consisted of about twenty young females dressed mostly in green grass skirts with yellow accents and red tops and with colourful lei-type frills. Some of the young women also wore strings of beads. On their arms, they bore red and white corsages and colourful ribbons on both their wrists and on their upper arms and yellow and red *maramars*, each graced by a red and white flower. They carried long thin sticks and formed themselves into two parallel lines. Their movements were carefully choreographed as they whorled, spun, and swayed, while clashing their long sticks against the sticks held by their opposite number arranged in a parallel line, all the while chanting in the local dialect.

A traditional Yapese Stick Dance troupe parades into the Seabee Base on Yap

The dancers form themselves into two parallel lines

The Dance begins . . .

Young Yapese learn their traditional dance movements at an early age

Dancers swirl and their sticks clash!

Spectators, both local and invited, watch as the dance performance continues

A Micronesian maiden adopts a commanding stance

Afterwards, the second group of young people took to the dance field to regale the spectators. The second group consisted of both young males and young females. The females wore the traditional Yap multi-coloured grass skirt of red, white, blue, yellow and indigo colours. Their skirts were layered at the waist so that two rows of short multi-coloured grass at the top of the skirt gave greater definition to their hips. In this group, the young women wore only colourful lei-type frills as well as strings of beads on their upper bodies. Their wristbands and upper arm bands were made from palm fronds. The male participants wore a plain skirt made from pandanus leaves worn over short pants. As with the female participants, the males also wore colourful lei-type frills and strings of beads. Both sexes wore *maramars* made from pandanus leaves, some bearing a floral decoration. This second group used more traditional thick bamboo sticks, which made a strong clacking noise when they were struck together. The choreography of the dance followed the same format as the first group as two parallel lines of dancers whorled about, clashing their bamboo sticks together while making a great shout or chant.

The second group of stick dancers prepares to perform at the Seabee Base

These young dancers of Yap were following the traditions of their forebears, as the local Yapese elders watched them perform. Throughout Micronesia Yap was regarded as having best preserved their Micronesian island culture and in the dance performances that day I could see the truth of those sentiments. Both groups of young people impressed all of the spectators with the colour and verve of their exotic dances.

The dancers raise their bamboo sticks at the beginning of the dance

The dancers dip and swirl . . .

Their sticks clash in a kaleidoscope of colours ...

Clack, Clack, as the sticks are struck against each other . . .

The colourful Chiru or Stick Dance of Yap reaches its conclusion

The hibiscus skirt worn by Yap women in the Stick Dance is dyed in bright colours

A colourful fringed collar is worn by the female dancers . . .

And a string of beads

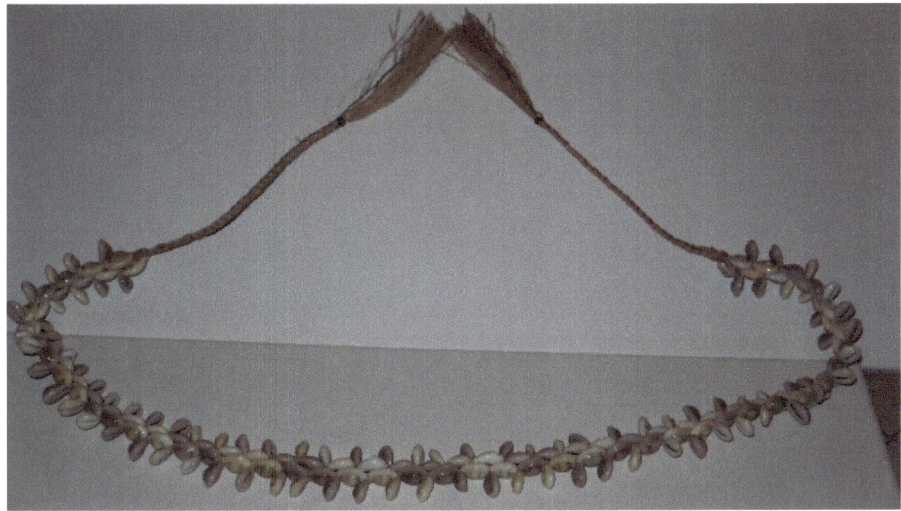

The cowrie shell belt is less commonly seen amongst dancers

A close-up of the intricately woven cowrie shell belt worn by some women in Yap State

After the Stick Dance performances concluded a great feast followed. A full pig was roasted and was displayed replete with a hard fruit in its mouth. In addition to pork we feasted on fish, taro (made into poi), breadfruit, sweet potato and a coconut pudding and fruits included papaya and banana, as well as other delights. The ceremony and feast at the Seabee base was the cultural high point of my visit to Yap Island.

Earlier during my stay on Yap, I had a conversation with the young female clerk at the Yap Women's Association about Yap customs and I was struck by her comment that she did not understand why people like me would come to her island because "there was nothing to see on Yap." I tried to explain to her that apart from its physical beauty Yap possessed a unique and charming culture and that I hoped the people of Yap would be able to preserve the best of their unique culture as the modern world increasingly reached out to Yap.

The time had now come for me to depart Yap and to continue my onward journey through the Federated States of Micronesia. The following day, with satisfaction at having been privileged to see a unique Pacific culture on the cusp of modernity, I returned to the ramshackle Yap aeroport and boarded an Air Mike flight to my next destination.

The vibrant colours of the hibiscus grass skirt worn by female stick dancers on Yap

NOTES

[1] It was quite common for western explorers to misunderstand the meaning of a local word and thus delude themselves into naming a place after something utilitarian, such as a paddle. My own country of Canada received its name when the early French explorers sought the name of the place they found themselves in and the indigenous people told them they were in "kanata", which meant "village".

7

TRUK LAGOON - THE SUNKEN MUSEUM

I began a long trek to my next destination, Pohnpei Island (formerly Ponape) the main island of the Federated States of Micronesia. That journey crossed most of the Federated States of Micronesia and involved stops on Yap, Guam, and Truk (now Chuuk) before finally reaching Pohnpei. So it was something of a 'milk run' in the air on board an Air Mike B-727.

Truk has a more unusual geological history than other islands in Micronesia. In a way, it masquerades as an atoll, with a 225-kilometre barrier reef encompassing a great lagoon that is 2104 square kilometres in area. At its widest point, the Truk Lagoon is 64 kilometres wide but has only five passages for ships to enter and depart. As a result, Truk Lagoon forms a natural sea fort. The Truk Lagoon is large enough to hold within it all of the islands that once formed the US Trust Territory of the Pacific. Given its size, a whole naval fleet could anchor at Truk and that is exactly what the Japanese did with Truk before and during World War Two when the Imperial Japanese Navy made Truk its main naval fleet base in the Pacific. But Truk is not actually an atoll. The great lagoon was once a large mountainous oceanic island before most of the island sunk beneath the waves. The remaining islands around the lagoon are merely the tops of mountains that remain from the single island that once rose high above the Pacific Ocean. Thus Truk is not regarded by geologists as a true atoll because the height of the remaining fringe islands differentiates it from typical atolls in which the land has eroded into the ocean leaving only low-lying palm-fringed spits of land surrounding an interior lagoon.

During World War Two, Truk Lagoon's status as the main Japanese naval base in the Central Pacific made it a tempting target for the United States Navy. On February 17, 1944, the US Navy mounted a massive air raid against the Japanese fleet at Truk called Operation Hailstone, which resulted in the sinking of forty Japanese warships and merchant ships in the relatively shallow waters of the lagoon. There the ships lay on the bottom in the years after the war. But by the Eighties, those ships had become encrusted with marine life and that made Truk a scuba diving paradise. Truk began to draw divers from all over the world. At Truk, Jim and Susan, my scuba diving acquaintances from Yap, left our Air Mike B-727 aeroplane to dive on the Japanese wrecks in the lagoon.

As at other island stops, the rear staircase on our B-727 was deployed from the fuselage so that passengers could embark or disembark from our B-727. The rear staircase is a unique feature of the B-727, and while our aeroplane sat on the tarmac I took advantage of the open staircase by asking the Air Mike cabin crew for permission to step off the aeroplane briefly so that I could take some photographs of Truk after the deplaning passengers had left. The Air Mike staff said 'Sure – go ahead, but don't be long because we will board passengers quickly.' So I descended the staircase in the rear of the aeroplane and stepped out onto the tarmac on Moen Island, the location of the air terminal, and I took a few quick photographs of Truk before any of the new passengers boarded. From the air terminal on Moen, I could see the peaks of some of the other islands of Truk Lagoon rising above the ocean in the distance.

At the airport on Truk Lagoon. Note the high islands in the distance, which are the remnants of high mountains that once loomed over the Pacific here.

Moen Island is the major island within Truk Lagoon and it hosts the airport, which is in the extreme North-East part of the island having a runway paralleling the coastline. The economy of the island is largely devoted to tourism, especially scuba-diving, and to copra. On some of the outer islands at Truk at the time of my way-stop visit in 1989, there was only a subsistence economy. The public sector economy on Truk, as on other islands in Micronesia, largely depended on transfers from the United States government. The lack of economic opportunities, along with the social dislocation brought about by an infusion of American cash and culture, were probably the reasons why the suicide rate in 1989 among young males between 15 and 24 on Truk was very high in comparison to other parts of the world. Public health and good sanitation also remained a continuing issue on Truk in the Eighties. As recently as 1982, seven years before my visit, there had been an outbreak of cholera on Truk.

Truk was only a transit stop for me and as new passengers crossed the tarmac to join my B727 flight it was time for me to re-board the aeroplane, and prepare for the final leg of that day's flight, which would take me to Pohnpei, the capital island of the Federated States of Micronesia.

8

POHNPEI - THE CAPITAL OF AN EMERGING ISLAND NATION

As my Air Mike 727 approached Pohnpei, an island that until a year or two before my visit had been known as Ponape, I spied a towering mountain rising to 780 metres, a natural monument befitting the largest island in all of eastern Micronesia. Pohnpei is both the major island and the capital island of the Federated States of Micronesia. It is a lush island with waterfalls and inviting pools, and it is sometimes called the garden island of Micronesia, an emerald gem in a blue sea, possessing thick jungles and mist-shrouded mountains surrounded by a barrier reef. The island is just above the equator at 7 degrees north latitude and it lies approximately half the distance between Honolulu and Manila. It is a rainy island, annually drenched with between 380 and 500 centimetres of rain along the coast and over 1000 centimetres of rain in the interior. Being close to the equator the climate is quite warm, with normal daytime highs hovering between 29 and 35 degrees Celsius. However, any discomfort due to hot temperatures is militated by the prevailing ocean breezes and the substantial cloud cover that normally lies over the island.

Pohnpei is the main island not only of the country but also of one of its internal states. There were 29000 people living in Pohnpei State when I visited in 1989, and of that number 90% lived on Pohnpei Island. Of the local population on the island, about 80% were indigenous to Pohnpei Island, with the remaining 20% consisting of out-islanders and Americans. At the time of my visit to Pohnpei, the national and state capital was the town of Kolonia, with a population of about 6300, according to a 1985 census.

The earliest peoples to arrive in Pohnpei were part of the great migration of Austronesian peoples that emerged from Asia via Taiwan and spread eastwards across the Pacific Ocean. That early migration may have made the initial inroads into Pohnpei by about 200 BC. Subsequently, there were invasions by foreign elements, firstly by the Saudeleurs who established an autocratic dynasty over the island, and then subsequently by the Isokelekel, or Idzikolkol dynasty, who were invaders from the island of Kosrae, an island which lies to the south-east of Pohnpei. The Isokelekel dynasty established a more decentralized and feudal government structure in place of the centralized Saudeleur dynasty, and that traditional form of governance was still in existence at the level of village life in Pohnpei at the time of my visit.

The earliest western contact with Pohnpei came intermittently from Spain, firstly in 1529, with the arrival of the navigator Álvaro de Saavedra, and then in 1595 by Pedro Fernandes de Queirós but neither of those navigators established any direct Spanish political control over the island. In the nineteenth century, Pohnpei began to be visited by western whaling ships during the heyday of the Pacific whaling industry. Service on a whaling vessel was often harsh and the crews were not made up of the best men. Consequently, a large western community of assorted deserters, castoffs, and riff-raff from the whaling ships took root in Pohnpei, although the indigenous people were not always receptive to the arrival of foreign residents. In 1828, a man named James F O'Connell, whose subsequent nickname became 'the tattooed Irishman,' landed on Pohnpei after his ship was wrecked on the reef surrounding the island. O'Connell narrowly avoided possible death at the hands of the indigenous people by dancing an Irish jig, which the indigenous Pohnpeians found to be hilarious. He continued to ingratiate himself with the local community by then courting and marrying the daughter of a local chief, after which he was tattooed according to island tradition and thus gained his nickname. In 1833 O'Connell left the island on another vessel and returned to America where he joined the PT Barnum circus to display his tattoos. He also wrote a book about his experiences in Pohnpei and his literary efforts helped to increase western awareness of the island.

By the mid-1850s Pohnpei was a major stopping port for the Pacific whaling fleets, with more than fifty vessels stopping there annually. Protestant missionaries from New England also became established on the island during this period through the so-called Morningstar missionary ships, and they began the process of converting the island to Congregationalist-denominated Christianity. During the US Civil War the frequency of whaler visits to Pohnpei by Union ships persuaded the master of the Confederate naval raider *CSS Shenandoah* to stop by in 1865 and to attack and sink several of the Union whaling ships that the *Shenandoah* found there.

The influence of the US government was first felt on the island in 1870 when the *USS Jamestown* stopped by and compelled the indigenous chiefs to enter into treaty relations with the USA and to permit foreigners to obtain title to land on the island. But it was the Spanish, having made the initial sovereignty claims to Micronesia as early as the 16th century, that ultimately established a nominal colonial rule over Pohnpei. Spain occupied Pohnpei in 1886, in the aftermath of the papal arbitration that awarded sovereignty over the Caroline Islands to Spain but also gave some commercial rights in the islands to Germany.

The indigenous people did not take kindly to Spanish colonial rule, with its forced labour requirements, nor did they accept Spanish proselytizing efforts that sought to establish a foothold for the Roman Catholic Church on the island, and indigenous uprisings resulted. The Spanish were forced to establish a defensive position at Kolonia, which the Spanish called Santiago de la Ascensión, and there they built Fort Alphonso XIII and the Spanish Wall to protect a governance compound. But following Spain's defeat in the Spanish-American War, and its subsequent loss to the United States of the Philippine Islands, and Guam, Spain's hold on the Carolines slipped, and Spain sold all of its remaining Pacific island colonies to Germany in 1899.

The remains of the Spanish Wall in Kolonia, Pohnpei, 1989

The Germans were commercially-minded colonizers and they made a determined effort to develop the copra industry in Pohnpei but like the Spanish, the Germans were frustrated by the indigenous people's refusal to submit to colonial rule. A very serious uprising broke out on the island in 1910-1911, the so-called Sokehs Rebellion, which is named after a prominent geographic feature near the capital town of Kolonia. The rebellion was sparked by the beating of a local man by a German overseer which enraged the locals, who then set upon the German and killed him. The Germans vowed revenge at this insult to white rule and took several months to assemble a suitable force from Germany's other Pacific possessions. Kolonia was blockaded, and eventually, native troops from the German Solomon Islands and Kaiser Wilhelm Land (North-East New Guinea) were landed at Kolonia along with sailors from the cruiser *SMS Emden* and they attacked the rebels, driving them up Sokehs Ridge where the rebels were defeated after a sharp engagement. The Germans took their revenge upon the rebels by executing seventeen leading rebels and exiling a further 426 supporters of the uprising from Pohnpei. After the battle, the Germans created two cemeteries in Kolonia, one for the white sailors and marines from the *Emden* and another for the German Melanesian troops killed in the action. As for the dead rebels, the Germans buried them all in a mass grave.

Sokehs Rock, the site of a battle between German forces and indigenous rebels in 1910

In 1914, World War One broke out and Japan swiftly joined the allied cause. The Japanese sent their navy into Micronesia to take possession of the German colonies. Pohnpei was the only place in Micronesia where there was any real attempt at German resistance. A junior official retreated into the interior jungles of the island with a detachment of local Micronesian troops but after a few days the hopelessness of the German cause became readily apparent, and the German official surrendered his small force to the Japanese. In the meantime, the Japanese had undertaken a house-to-house search for guns and ammunition but despite their position as the conqueror, the Japanese took great care to re-pack everything they had disturbed within each house.

After the war, the Japanese acquired Pohnpei as part of their South Seas Mandate from the League of Nations, which marked the start of substantial commercial and agricultural development on the island. The Japanese also fostered emigration from Japan and by 1941, when World War II began in the Pacific Ocean there were 14000 Japanese and Koreans[1] living on Pohnpei and only 5000 Micronesians. Although the Japanese encouraged intermarriage with the indigenous population such unions were relatively uncommon. During the war Pohnpei was bombed by the United States and Kolonia was severely damaged but the island was not invaded by US troops. After the war, the United States took possession of the island and all Japanese settlers across Micronesia, including Pohnpei, were rounded up and sent back to Japan, including those who had entered into marriages with an indigenous person, and who had children with them, thus leading to the forcible breaking up some families.

The United States was granted a strategic trusteeship mandate over Micronesia by the United Nations after the war and Pohnpei and the other Micronesian islands largely disappeared from the world behind the curtain drawn across them by the United States. In the Sixties, however, the growing decolonization movement across Asia and Africa prompted the United States to begin to actively meet its obligations to the islands as a UN mandatory power, and the people of Micronesia began to send delegates to a US creation, the Congress of Micronesia. As time wore on it became clear that ethnic and cultural differences between the various island groups in Micronesia meant that many of the islands would go their own separate ways, and there would be no United States of Micronesia. Nor would there be a wholesale desire on the part of most Micronesians to join the United States. Most of the Caroline Islands decided to form a single state and a constitution for the Federated States of Micronesia that was drafted in 1975, and adopted in 1978, came into force on May 10, 1979. By a formal proclamation of US President Ronald Reagan, the Trust Territory of the Pacific was terminated insofar as it affected the Federated States of Micronesia on November 3, 1986, which effectively granted the country its independence. The United Nations, however, did not formally recognize the termination of the trust until 1990 by which time it was a *fait accompli*.

Pohnpei joined with all of the other Caroline Islands with the exception of Palau to form the Federated States of Micronesia, which then signed a Compact of Free Association with the United States that gave the US the right to base military forces in the country in return for an obligation by the US to defend the country if it was attacked. Under the terms of the Compact, the US additionally agreed to support airport safety, meteorological services, public health, and emergency relief. The US also promised to continue to subsidize the Federated States of Micronesia because much of the formal economy consisted of government-related public services financed by the United States. Beyond the public services, economic activity on Pohnpei was largely limited to subsistence farming, copra production, and a few pepper plantations that exported gourmet pepper to the USA. Tourism remained a nascent industry.

The new country encompassed 700 square kilometres of land scattered over 2.5 million square kilometres of the Pacific Ocean, with about half of the land area on the island of Pohnpei. The population of Pohnpei was 30,819, in the late Eighties, and half the population in the entire FSM was under fifteen years of age, making the new country one of the youngest in the world. The FSM was organized into four separate states ranging, from east to west, Kosrae State, Pohnpei State, Truk State, and Yap State. The national capital in 1989, Kolonia on Pohnpei, was also the capital town of Pohnpei State. When I visited the FSM back in those pre-internet days there were no private newspapers, and all of the existing newspapers in the country were owned by the government. However, the privately-owned *Pacific Daily News* which was published out of US-owned Guam was available in the FSM.

The new country was still feeling its way onto the world stage in 1989 when I visited the FSM. On December 16, 1988, only a few weeks before my visit to the country, Japan and the FSM exchanged diplomatic letters resulting in Japanese recognition of the sovereignty of the FSM and providing for the establishment of diplomatic relations. Somewhat wryly, the Japanese diplomat present for the ceremony noted that Japan and the Federated States of Micronesia, "have historically enjoyed the tradition of good relationship."

The government newspaper[2] also noted that the country had joined the Asian and Pacific Coconut Community (APCC) through the FSM Coconut Development Authority in hopes of further developing its copra industry. With a view to encouraging the growth of tourism to Micronesia, the legislature had also passed a bill to authorize the payment of the membership fee for the island nation to join the Montreal-based International Civil Aviation Organization. Independence brought new international challenges as well, with the government newspaper noting that fishing talks with Taiwan had broken down over the FSM's worries about the sustainability of its fishery. To buttress its maritime territorial claims the newspaper reported that the President of the Federated States had signed a law to establish an exclusive economic zone around the Federated States of Micronesia, and to expand the size of the country's territorial sea.

The Pohnpei aeroport on Takatik Island where I arrived on the island was a modern aeroport terminal, and I quickly located my hotel-provided transfer service and departed for my accommodations at The Village hotel. A causeway connected the small island on which the aeroport was located with the main island of Pohnpei. I found accommodations at The Village to be quite charming, boasting twenty-one thatch-roofed cottage bungalows that were set into a lush hillside providing views of the distant ocean. The rooms were sited in such a way as to take advantage of both scenic views of the reefs and lagoon through the lush foliage, as well as to receive the cooling effects of the prevailing winds off of the ocean.

The Village was located near the small village of Awak, about eight kilometres east of Kolonia. Hotel entrepreneurs built The Village in 1976, hoping to create an ambiance that would fit into the natural beauty of the island. In that, they succeeded, as the hotel consisted of separate cottages built of locally-sourced materials and designed to look like the hut of a local chief. While the exterior was very traditional, the interiors were quite modern with a large waterbed, indoor bathrooms with hot water, electricity, overhead fans, and wicker furniture.

A thatched roof overlook at The Village on Pohnpei, 1989

My room was quite lovely, and I revelled in the local ambiance. At The Village, it seemed that I could truly feel like I was living in the South Seas. A sudden shower only added to the island ambiance as I unpacked. But I was perplexed by two large sheets hung, hammock style, over my bed from the thatched roof ceiling. Once I was settled in and embarked upon an exploration of the hotel and its suspended walkways I asked the hotel management about the purpose of the large sheets suspended above my bed. The hotel staff explained that sometimes at night small native geckos infiltrated the guest cottages and a number of guests had been surprised from their sleep when geckos had lost their footing and landed on them. Some of the guests, especially female guests, the staff said, had expressed great discomfort at being awakened in the night by the sudden arrival of a gecko landing on their sleeping bodies! To prevent any unfortunate landings in the night the suspended sheets were installed as a sort of gecko safety net, much like the nets for circus trapeze artists. Once installed, the gecko complaints ended.

The pathway leading to my thatched cottage at The Village

The interior view of my cottage at The Village shows the protection above the bed for geckos that lose their grip in the night!

The Village also boasted a pub and restaurant that was built open-sided and was said to be the largest thatched roof building in Micronesia at that time. It was named after the nineteenth-century Irish-American beachcomber James F O'Connell, and called *The Tattooed Irishman*. Service at the hotel was excellent, and I learned that the proprietors had over-hired recognizing that many of the local people worked on island time, and therefore often took unexpected days off.

After exploring the hotel I went into Kolonia and examined the town. Kolonia was then the capital town of the Federated States of Micronesia, although a new capital at Palikir, in the interior of the island, was nearing completion and would soon replace Kolonia as the national capital. According to the government media, the planned occupancy of the new capitol complex at Palikir had just been postponed from February, 1989, to May, or June, 1989, to allow additional time to ensure that everything was ready for the move of the government to the new capital.

Kolonia was a town of ramshackle shops, warehouses, a co-op, and a farmers market. In my mind it appeared much as an old-fashioned South Seas seaport would have appeared in former days. In Kolonia, I saw the old Catholic Mission Bell Tower and all that remained of an old German church that had been demolished by the Japanese, as well as the remnants of the old Spanish Wall that was built as protection for Fort Alfonso XIII in 1887. I also viewed from a distance the famous Sokehs Rock, the outcropping cliff face that was the site of the 1910 rebellion against German rule.

I wandered through a village on the outskirts of Kolonia called Kapingamarangi, which was the abode of Polynesians who came to Pohnpei from Kapingamarangi and Nukuoro atolls after typhoons and famines had displaced them from their home. The isolated atolls were the home of the only Polynesians in the FSM. The displaced Polynesians had built open thatched homes on platforms and lived a more traditional and much poorer lifestyle than the local Micronesians.

Through its long history of western contact, and its current role as the capital island of the new nation of Micronesia, Pohnpei had been modernized to a considerable degree, including in its sartorial customs. In Kolonia, the people mainly dressed western style. The local men generally wore pants either short or long, and a western-style shirt usually with a collar. I saw no males wearing traditional loincloths except at the Pohnpei Cultural Centre. Topless females were part of the island's cultural traditions but not seen in Kolonia, where western-style fashions or lava-lavas paired with a blouse were generally worn by the local women. Some women in the rural areas of the island still went about topless, as did the women displaying traditional crafts at the Pohnpei Cultural Centre. However, as on Yap Island, which I had visited earlier, there was a strong custom in Pohnpei that females must not display their thighs in public. It was a custom that was not always respected by western female tourists. As I stood on a street in Kolonia one day I saw two local teenage boys loitering around the stairs that led to the doorway of a store as a young western couple approached. The young western woman was wearing very short shorts that exposed most of her thighs, and the Micronesian boys froze as she approached and then immediately turned to stare at this rare sight of bare thighs as she passed them on her way into the store. Generally, the local women in Micronesia severely judged western women who insisted on displaying their bare upper legs in public, which usually discouraged even the most obtuse western tourist from offending local sensibilities.

I ended my walkabout in Kolonia and returned to The Village. A tour guide book that I was using during my Micronesian travels told me that local taxis in Kolonia were merely shared pick-up trucks and therefore taxis were not easily distinguishable from other private vehicles. To hail a taxi the guidebook said to just wave down any pick-up truck that looked to be crowded with people if I wanted a ride. So that is what I did and I waved down a taxi filled with several people and told the female driver where I wanted to go. She told me to hop in the back of the truck. In the bed of pick-up, I found myself with several charming young people and children who seemed amused by my presence, and who smiled shyly and stared at me. By this time I had begun to realize that this probably was not a taxi and that was confirmed when the woman dropped me near the entrance to The Village and refused to take any money for the ride. I had mistakenly flagged down a large family returning from the town to the countryside but it was a nice example of the local friendliness that I experienced from all of the Micronesians that I met in Pohnpei.

One day at *The Tattooed Irishman* at The Village I met two American businessmen who were also staying at the hotel. Both of them were from California. David was President of an international company involved in import/export trade in the islands, a sort of modern version of a traditional South Seas trader. He was in Pohnpei to purchase gourmet pepper in bulk for the American grocery market. Pohnpei pepper was renowned in the American marketplace at the time for its flavour, and I myself purchased a few packets of Pohnpei pepper to take home with me as gifts for relatives and friends.

Darrell was the other American businessman and he was the president of a pharmaceutical consultancy business. Darrell had become involved in a business dispute in Pohnpei that had led to litigation in the local courts and he had come out to Pohnpei to prepare for an upcoming trial. David told me that some of the locals in the Pohnpei business community were impressed by the fact that Darrell had come out from the United States for the forthcoming trial, rather than merely instructing his lawyer to represent him in court. As a lawyer myself I was very interested in the legal system in this country and I learned that it was largely based on the US judicial system and US legal practice. The judges were appointed by the President of the Federated States of Micronesia, upon the advice and consent of the unicameral Congress of the Federated States. There were two judges appointed to the Supreme Court of Micronesia and in 1989 both of them were Americans who had obtained some legal experience in the islands before being elevated to the bench. The Chief Justice, who was based in Pohnpei, had formerly been the Attorney General of the Federated States of Micronesia, while the other judge was based in Truk.

One day Darrell and I drove out to have a look around the new capital at Palikir, which would shortly replace Kolonia as the capital town of the Federated States of Micronesia. On the drive out to Palikir, we passed through areas of lush jungle as well as some more open areas. Darrell had previously told me that he had served in the US Army Special Forces during the Vietnam War and travelling through the lush countryside of Pohnpei brought back some painful memories of Vietnam for him. As we passed a part of the road where the jungle came down to the road on both sides his mind went back to his experiences in Vietnam and he remarked; "This is where they [i.e. insurgents] usually wait to get you – where there is a lot of vegetation on either side of the road and you can't see them. Of course, they know that you might be expecting an ambush there, so sometimes they will wait until you pass back into the open area and then they surprise you there."

The new capital complex at Palikir was a very modern government campus. It consisted of nine one or two-story concrete buildings connected by walkways (some of them covered) and it had been under construction since 1987. Now, in early 1989, it was virtually complete and only the finishing touches were being made to it. We wandered around the almost deserted complex while a few labourers were making final installations. The national government had been scheduled to move into the complex in February, 1989, but circumstances had delayed the government's move from Kolonia until May or June. The new government capital project had also involved the construction of the seven-kilometre paved road from Kolonia to Palikir that we had just driven over to visit the new capital town.

Two views of the new capitol complex under construction at Palikir in early 1989

The Village was almost full of guests during the time that I was there and in addition to the American businessmen whose acquaintances I had made, there was also a group of boisterous young scuba divers from Nevada who were exploring the local reef. Two of them were a young married couple, who were there on a diving honeymoon. Scuba diving at Pohnpei was good although perhaps not quite as good as the diving at Palau for its sheer quantity of visible ocean life, or Truk, for the number of historic wrecks in the Truk lagoon.

One night while I was sitting in *The Tattooed Irishman* having a drink I struck up a conversation with a very young married couple. The man was a handsome young indigenous Micronesian, and I suspected that he must belong to a wealthier or politically-connected family. His young wife was a young American and both of them were in their very early twenties. They had met when the young woman had come to Pohnpei on a Peace Corps mission. They made a fine-looking couple, he a handsome dark-complexioned young man, and his wife a beautiful blue-eyed blonde woman. Throughout the period of western contact there had been untold numbers of unions between Caucasian men and indigenous women but here was a rare example of the reverse, a Caucasian woman and an indigenous man. And from their behaviours towards each other, they were obviously in the throes of young love. I wished them well s they left but an older American man sitting nearby at the bar who had been observing our conversation looked at me as the two young lovers walked out and remarked "That girl must've broken her father's heart," a display of a harsh American attitude towards race, and interracial marriages, that was still prevalent in the Eighties.

The man at the bar was also critical of the Peace Corps, and he declared that there was a notable propensity for young American Peace Corps volunteers on isolated islands to 'go native.' He told me of a friend's wife who had worked in a dental office in Guam where she had frequently served Peace Corps volunteers on assignments in Micronesia. She would often see Peace Corps volunteers arriving from isolated locations in Micronesia who were dirty and unkempt in appearance, even wearing mismatched footwear. But I felt compelled to remark to him that although I had not gone out to the most remote islands the Peace Corps volunteers that I had met on Yap had left a very positive impression with me.

On another day I decided to participate in an organized jungle trek with a group from the hotel to the famed Kepirohi Waterfall. We drove east of The Village in the hotel van beyond Awak Village to the starting point of an unmarked and rough trail leading up to the falls. It was a short 45-minute hike to Kepirohi Waterfall and the trail was not too difficult, although the humidity in the interior of the island was high, and I was sweating profusely by the time we reached the end of the trail. We were rewarded by the sight of an inspiring waterfall. The Kepirohi Waterfall is 21 metres high and is quite impressive. There were some quiet pools below the waterfall in which we could swim, or wade, to cool off but after a couple of people in our group reported a strange nipping feeling on their legs our guide laughingly admitted that there were freshwater eels living in the pools beneath the falls. Although the eels posed no real threat to people, a few members of our group did experience a little nibbling!

Trekking through a jungle to Kepirohi Waterfall

On Pohnpei, the local diet is heavy on fish and other sea creatures, and yellowfin tuna, dogtooth tuna, mahi-mahi, jack, bonito, barracuda, snapper, grouper, wrasse, and porgy can all be caught in Micronesian waters. Onshore the people of the island cultivated yams, which formed a staple part of their diet, as well as breadfruit. Coconuts and other various fruits were widely available.

To learn more about the cultural customs of Pohnpei, I joined a group visiting the Pohnpei Cultural Centre where the traditions of the indigenous people of the island are kept alive in the face of onrushing modernity. We were greeted with the presentation of floral coronets and then we watched as a group of people under the direction of elders displayed some of the traditional arts of the Pohnpeians. First, we watched a traditional palm leaf weaving demonstration by young women, and then some of the young men demonstrated coconut husking by implanting a sharpened branch in the ground and then impaling a coconut over the sharpened point of the branch at the weakest point of the nut, pressing down hard to split the coconut. The husks were later used as vessels in which to serve a drink of *Sakau* to the visitors.

Visitors to the Pohnpei Cultural Centre receive a floral coronet

A Palm-weaving demonstration at the Cultural Centre

The Pohnpeians are well-known in Micronesia for their preparation of the unique drink called *Sakau*. Only in Pohnpei do Micronesians manufacture this drink, which is both a traditional ceremonial drink and one that is also served as a staple in the local taverns. At the cultural centre we watched as this drink was prepared for us. *Sakau* is made from the roots of a pepper plant, *piper methysticum*, and the root is first pounded to a pulp on a flat rock and then water is added to it after which the mixture is wrung through a strainer made from hibiscus bark. The now strained liquid is then poured into a coconut husk and served. In appearance, *Sakau* looks like a muddy milkshake and I found the texture to be somewhat slimy as one drinks it. *Sakau* is a mild narcotic and the effects are soon felt. At first, my lips and tongue began to feel numb, and then the numbness spread throughout my body giving me a soothing and calming feeling. However, the mild narcotic effect is not long-lasting and soon dissipates. Although it is made from a different plant than *Kava*, a similar drink prepared on Polynesian islands, the effects on the mind and body are similar.

The root of the pepper plant is pounded into a pulp as a first step to making Sakau

Water is added and the pulp mixture is strained through hibiscus bark

The mixture is allowed to settle before the Sakau is served

In the meantime, a stick is sharpened to husk coconuts

Coconuts are husked while a visitor looks on. The husks are later used as serving vessels for the Sakau.

Demonstrations at the Culture Centre end with a traditional chant

After a day at the Pohnpei Cultural Centre, I returned to The Village to prepare for the highlight of my journey to Pohnpei, a visit to Nan Madol, the ancient stone city that has been called the Venice of the Pacific.

NOTES

[1] At that time Korea was part of the Japanese Empire.

[2] *The National Union* Vol. 10, No. 1, Kolonia, Pohnpei, January 1989

THE MYSTERIOUS CITY OF NAN MADOL

The highlight of my journey to Pohnpei was undoubtedly my visit to the City of Nan Madol, the ancient stone city of the Saudeleur dynasty that has been called the Venice of the Pacific. Situated on the east side of Pohnpei on a reef adjacent to the small offshore island of Temwen, an ancient stone city rises from the Pacific Ocean. This is Nan Madol, the fortress city of the little-known Saudeleur Dynasty that ruled Pohnpei for 500 years. Today it is but a ruin but once it must have been a striking sight to anyone who approached it by sea.

Trip-n-tour had arranged for my visit to Nan Madol, which began at dockside in Kolonia where I boarded a small motorboat with three other people. Soon we were headed off from the shore and were skirting the little islets that populate the reef around Pohnpei. As our journey unfolded I became acquainted with some of my fellow passengers. Two of the passengers on my boat were Americans, a man who was a pilot for Continental Airlines and his girlfriend, who held an administrative job back in the United States. The pilot told me that the couple was visiting Pohnpei because he loved Micronesia and he had never missed a chance to visit these islands and now wanted to show the islands to his girlfriend. Although the Continental pilot was American-born his girlfriend hailed from Iran but she carefully described herself as a Persian. In the Eighties, many Americans still harboured bitterness towards Iran for the kidnapping and holding of American diplomats hostage in Tehran in 1980, and that was probably the reason why she was cautious about identifying her origins. My Persian acquaintance was clearly one of those who had escaped from the harsh conditions imposed on the Iranian population by the victorious mullahs, following the 1979 revolution against the Iranian monarchy.

We passed small islets on the reef surrounding Pohnpei on the passage to Nan Madol

Pohnpeians in 1989 still travelled on the sea in outrigger canoes like their forefathers

After a time our boat began turning towards shore and I awaited with anticipation the sight of the magnificent stone ruins of this mysterious city that I had read about. Slowly the ruins of the city began to appear, seeming to float upon the sea. Nan Madol was massive, a city built of almost unmoveable basalt logs that still makes archaeologists puzzle over the ingenuity of the early inhabitants in transporting and raising such massive stonework. Nan Madol is the only city in the world that has been built upon a coral reef and it is still a matter of archaeological conjecture how the material for the massive stone ramparts was moved to Nan Madol and raised into position by a culture that had only limited technological capabilities. There are ninety-two islets at Nan Madol, and they are formed into a modern ruin of basalt log walls, paths, and canals. The most prominent archaeological feature is the ramparts of the temple and burial vault of Nan Dowas, which boasts walls that are eight metres high.

The walls of Nan Dowas

Nan Madol was designated as a US National Landmark when Pohnpei was under US administration as part of the Trust Territory of the Pacific Islands. For half a millennium it served as the fortress city, religious centre, marketplace, and royal capital of the Saudeleur monarchy. The city was built sometime between 1100 AD and 1200 AD by two brothers, Olisihpa and Olosohpa who came to Pohnpei with their followers from across the sea and settled on the island. After a few false starts at building a new city, the two brothers decided that the best site for their capital was on the reef at Temwen, supposedly because there was evidence of an earlier but now submerged city of the gods that lay just offshore from the proposed site of Nan Madol.

The legends of Pohnpei state that Olosihpa and Olosohpa were great sorcerers, or magicians, who derived their power from their gods and they used magic to build their city. After settling on Pohnpei both brothers married Pohnpeian women and both began aristocratic lines but after the death of one of the brothers the other brother declared himself to be a King, and the monarchical line of the Saudeleur dynasty was established at Nan Madol.

The purpose of Nan Madol was to create a city that separated the nobility from the commoners of Pohnpei, and the city represents the earliest example of a centralized political entity in the western Pacific. Within the sacred complex of Nan Madol, there were temples, altars, and oracles. The city held a population of about 1000 people who were bound by a rigid hierarchy in which the commoners were required to serve the nobility. Tribute was also required from the more widespread island population of Pohnpei. For centuries the political structure imposed by the Saudeleur monarchy remained stable. But as the long years passed the Saudeleurs became increasingly oppressive towards the people of Pohnpei, and the tribute they demanded from the Pohnpeians became increasingly onerous. Political stability became fragile. Sensing an opportunity, a warrior from the island of Kosrae, which lies to the south-east of Pohnpei, mounted an invasion of the island, and that invasion also served to incite a local rebellion by the Pohnpeians that swept away the Saudeleur dynasty in 1628. The victorious Kosraean warrior, Isokelekel, established a new feudal political structure over the island, a structure that still exists on the island in respect of traditional matters today. However, over time, following the overthrow of the Saudeleurs, the importance of Nan Madol diminished, and sometime in the late eighteenth century, the city was finally abandoned.

The walls of the city of Nan Madol

Long abandoned, the city was overtaken by the jungle in 1989

Although it is unclear when exactly the city was abandoned it was definitely deserted by the time that Christian missionaries arrived on the island in the early nineteenth century. Missionary records dating from the mid-nineteenth century state that at that time the missionaries arrived in Pohnpei only elderly Pohnpeians could still remember a time when Nan Madol was populated, so the abandonment of the city must have been relatively recent from a historical perspective.

It is believed that basalt rock created through volcanic action naturally cooled into hexagonal shapes on the island. After extraction, archaeologists surmise that hexagonal basalt logs that were used to construct Nan Madol were moved, perhaps by rafts, to the site of Nan Madol and then somehow raised into position on the reef. The extraordinary effort that would have been required to undertake the construction of Nan Madol becomes quite apparent when one considers that some of the basalt logs were more than 7 metres in length and weighed up to 45 metric tonnes. The basalt logs were stacked criss-cross using the header and stretcher manner of construction. Once the walls had been built up high out of the water on four sides the interior of the resulting square, or rectangle, was filled in with coral rubble and soil to create an artificial islet. The walls of the remaining ruins in 1989 reached a height of about eight metres.

The giant basalt logs weigh many tonnes

An intricate system of canals connected the 92 inhabited islets on the reef at Nan Madol

The city was divided into two parts. Madol Powe, the upper part of the city, was its ceremonial centre and it included the great temple of Nan Dowas, which was made of basalt logs of up to one metre in diameter and several metres in length. Nan Dowas served as both the main temple and the mausoleum of the Saudeleur Dynasty. Madol Pah, the lower part of the city, was its secular centre and in that part of the city, stone platforms were constructed upon which pole and thatch structures were raised in the traditional manner of the Pohnpeians to house the city's population. The lower town was not wholly secular, however, as there was also an important temple in this part of the city, Nan Kieil Mwah (the Temple of the Good Lizard). Throughout the city, canals were cut through the reef to afford access to the various islets that contained some 130 separate structures. Tunnels beneath the islets connected the ocean to ponds located on some of the islets and on one of the islets called Darong, or Idehd, there was said to have been a large pool in the centre of the islet that held a sacred eel, which the Saudeleurs fed with turtle meat.

The city proper occupied an area of 28 square kilometres but the whole metropolitan area is actually larger. Pohnpeian oral history named the entire area Deleur and it included the city (called Kahnihmw), all of Temwen Island, the reef islets south and east of Nan Madol, as well as a northern district (called Metipw), and some portions of the east coast of Pohnpei proper (called Tamwerahi or Lepweltik) for a total metropolitan area of 49 square kilometres. Within that area, there are a total of 90 artificially constructed islets connected by an extensive canal network that led early observers to refer to Nan Madol as the Venice of the Pacific. Most of the canals were overgrown by mangroves and silted up when I visited in 1989, but the main channel to Nan Dowas, which is called *Dauen nan Kieil Mwahu* or the Channel of the Good Lizard, could still be navigated by motorboats at high tide.

Many canals are now overgrown and impassable

Many of the structures at Nan Madol had now collapsed by the time of my visit in 1989 but the overall site was still very impressive and it still constitutes the most significant monumental ruins remaining in the Pacific Ocean. Many of the indigenous people of Pohnpei harboured superstitions about Nan Madol when I visited the site in 1989, and felt that it should be left undisturbed. Many of the locals avoided this ancient site, fearing that it is an abode of spirits. According to a book written a few years after my visit, there are a number of legends that surround Nan Madol which suggest that it formed part of a gateway to a legendary lost continent of the Pacific.[1]

The first mention of Nan Madol by western explorers occurred in 1836, by James F O'Connell, the famed Tattooed Irishman, who said that the city was a ruin. Subsequently, a missionary in Pohnpei named Gulick reported in 1857 that the oldest Pohnpeians could still remember a time when the whole city was densely populated. In 1874, Kubary, an agent of a German museum, prepared the first good site plan for Nan Madol, which was later revised in 1896.

The Spanish seem to have taken little notice of Nan Madol during the period in which they controlled the Caroline Islands but the Germans who succeeded them took a much greater interest in it. Between 1905, and 1907, a German Governor, Berg, undertook extensive excavations at Nan Madol but Berg apparently violated a taboo when he broke into the tomb of *Pein Kitel*, which was said by the locals to be protected by the spirits of the dead. Astonishingly, Berg died mysteriously on the very day that he excavated *Pein Kitel*. In 1910 another expert from Germany named Hambruch undertook ethnological surveys and he obtained further information about the customs and traditions once associated with Nan Madol from the indigenous population. He also drew up the best map of the ruins to date.[2]

After 1914, the Japanese held possession of Micronesia and Japanese archaeologists performed extensive fieldwork at Nan Madol. However, we know little of their work because their work was not subsequently published, although Von Daniken, a writer of alternative history whose works were popular in the latter part of the twentieth century, apparently stated that Japanese archaeologists had recovered Saudeleur royalty entombed in platinum coffins within a submerged tomb at Nan Madol![3] There is, of course, no substantive evidence for that but it is also true that very little is known about the work of Japanese archaeologists at Nan Madol in the interwar years. However, the Japanese did undertake efforts to remove the vegetation that was slowly strangling the ruins, and, fortunately for posterity, they removed the banyan trees that were threatening to undermine the structural integrity of the remaining ruins.

After the United States assumed control of Pohnpei there were expeditions by the Smithsonian Institution (1963), which did not publish its results in full, and by the National Geographic Society in 1984, which likewise did not publish its full results. Some archaeological survey work by two archaeologists named Ayres and Saxe on behalf of the Trust Territory government was undertaken just prior to the FSM gaining its independence and has been made available, but much still remains unknown about Nan Madol.

Several important archaeological questions remain. Nan Madol was an artificially constructed stone city that consisted of approximately 226 million metric tons of basalt rock. All of that rock was cut and transported to Nan Madol where it was sunk into place on the reef and then built up to heights of about eight metres above the surface of the ocean. How was it possible for a Pohnpeian culture with limited technology to undertake and complete such a vast and complex engineering project? When the local people were questioned about it they claimed that the two brothers who founded the Saudeleur Dynasty, Olosihpa and Olosohpa, used magic to move and raise the stone logs. According to the Pohnpeian legends, the logs were quarried on the west side of Pohnpei and then floated through the air by Olosihpa and Olosohpa, who sometimes rode through the air on the logs to guide them! That has led one modern writer to suggest that it might be possible to levitate the logs by sound if the basalt logs, which are crystalline and magnetic, could be tuned to resonate at the frequency of gravity, and thus lose their weight. That would explain indigenous legends that describe the logs being spun upwards and to the east because if it were possible to levitate the giant stone logs the Earth's rotation would appear to move them eastwards once they were levitated.[4] Despite that intriguing suggestion, the reality is likely to be much more prosaic. The basalt rock was probably cut out and then rafted to Nan Madol where the logs were landed on the reef, hauled into place, sunk into position, or lifted up to form the walls all through the application of brute human force. Still, it remains a mystery how enormous logs and rocks that weighed up to 45 metric tons could have been manhandled into place. It hardly seems possible that such heavy objects could have been transported and placed through brute human force, even if the Pohnpeians had used inclined planes, well lubricated with coconut oil, to move the heavy stone logs upwards to set them into place. However, until further evidence comes to light, that remains the most logical explanation of how it was done.

As to where the technology to construct Nan Madol came from one possible explanation is provided in the legends of the island of Kosrae, the island which launched Isokelekel on his invasion of Nan Madol, and where large stone ruins are also present, although much smaller in scale than those at Nan Madol. On Kosrae, the legends speak of a ship from a foreign land that came from the northwest and brought skilled engineers who raised walls on land around the town of Lelu on Kosrae. It has been speculated that the ship may have come from Japan and that the early Japanese would have had the skill to build the walls found on Kosrae, as the ancient walls around the Imperial Palace complex in Tokyo would suggest. Nevertheless, even for Japanese artisans, the challenges of building a city on the scale of Nan Madol would have been of a wholly different magnitude from the building of the walls of the Imperial compound in Tokyo.

In addition to the artificial islets that have been constructed at Nan Madol, the builders also had to cut canals through the reef to connect the various structures built on the artificial islets (the actual translation of Nan Madol is 'the spaces between,' which is a possible reference to the canal system). According to the legends of Pohnpei the canals were built by a dragon that cut through the reef to complete the canals. It is striking that the indigenous legends of Pohnpei do not speak of the use of brute human effort by hundreds, or thousands, of people to build the city but instead suggest it was largely built by magic through the flotation of stone, and the use of a dragon to cut the canals. Such legends give rise to suggestions that people in an early stage of development may have witnessed advanced technology which they described as 'magic' or 'dragons' because they could not otherwise process what they had seen within their version of reality. That has given rise to many contemporary speculations about advanced races, or even extraterritorial visitors, but probably the truth has simply been forgotten or collectively suppressed because the reality was too awful to remember.

The indigenous legends of Pohnpei also apparently speak of an earlier city of the gods, the remains of which are now said to lie in the ocean offshore from Nan Madol, and possibly of yet another city even farther away under the water.[5] During the Trust Territory government's efforts to survey the site of Nan Madol in the early Eighties one of its consultants, Saxe, undertook some underwater work where he discovered a path marked by boulders and two columns underwater. Since then more columns have been suggested to exist by other underwater explorers. The columns seem to only be found underwater, and they are not present on land at Nan Madol, or elsewhere on the island.[6]

As I wandered around the ruins of Nan Madol I could not help but be impressed by the grandeur of this fallen city and I marvelled at the ability of the ancient Pohnpeians to build this city of stone with very little technology. I pondered how this great ruin could have been constructed with the tools available a millennium ago. As our guide acknowledged, experts in 1989, did not know how the city was constructed given the size and weight of the basalt logs that form the structures placed on the artificial islets. The Pohnpeians say it was magic, and feel uneasy about the city. Even in 1989, the indigenous population largely stayed away from the abandoned site, still believing that it was the abode of ghosts, and emphatically stating that at night strange lights were often seen moving through the ghostly city. Of that, I can say nothing because well before the sun began to set we boarded our boats and returned to Kolonia, where I spent a final night at The Village.

By now my journey to Pohnpei had reached its end and it was time to leave Pohnpei to journey to another part of Micronesia, the island of Palau at the far western end of the Caroline Islands. As the transfer bus drove through the countryside with its windows open the sounds of rural life in Pohnpei came to me. Heartrending screams told me that a pig was being slaughtered nearby for an evening meal. That is a scene that occurs as part of life on many islands in the Pacific, and elsewhere, but one that is kept hidden from modern urban dwellers in North America. We passed a group of smiling young schoolgirls dressed in their school uniforms of blue skirts and white blouses walking along the side of the road. As we drove past they all smiled, waved, and shouted greetings to us. That was a charming moment in a South Seas idyll.

Soon I found myself once more aboard Air Mike's milk run flight across the FSM on my way to Guam, enroute to Palau. I sat beside a young Japanese woman on the segment between Truk and Guam and I was delighted to learn that she had once visited Canada. But I was surprised when she told me that the part of Canada she had visited was Prince Edward Island, our smallest province, which lies off of the east coast. Somewhat puzzled as to why she would visit Prince Edward Island (which is admittedly a charming part of the country) I asked her why she went there and I learned that she had gone there because the island is the setting for the famous stories of Ann of Green Gables. She told me that Ann of Green Gables has a tremendous following in Japan and that many Japanese girls and young women have read the novels written by Lucy Maud Montgomery about the adventures of the headstrong redhead Anne Shirley that were set in Prince Edward Island in the late nineteenth century. Although I had travelled halfway across the Pacific Ocean to gain knowledge about Micronesia, I had rather surprisingly learned that stories from my own country had spread across the Pacific.

NOTES

[1] Childress, David Hatcher, *Ancient Micronesia & The Lost City of Nan Madol, including Palau, Yap, Kosrae, Chuuk & The Marianas*, Adventures Unlimited Press, Kempton (Illinois), 1998.

[2] National Park Service, *National Register of Historic Places Inventory – Nomination Form*, December, 1984, pages 3-4

[3] Childress, page 33

[4] Childress, page 53

[5] Childress, page 47

[6] *Ibid*

10

PALAU - THE LAST TRUST TERRITORY

My flight to Palau on Air Mike from Guam was uneventful except for the fact that the captain on the B-727 jet was a woman. In the Eighties that was still quite unusual, especially on a large passenger jet, and despite my extensive travels that was the first time I could recall a woman in command of an aeroplane in which I was a passenger.

The Republic of Palau (also traditionally called Belau) lies southwest of Guam, east of the Philippines, and north of Papua New Guinea, at the far western end of the Caroline Islands chain. Palau consists of some 200 volcanic and limestone islands ranging over an area of 650 kilometres from north to south and all but three of those islands lie within the protection of a barrier reef. Palau is very close to the equator, being only seven degrees and thirty minutes north of the line with the result that Palau has a very sunny and hot climate with daytime temperatures averaging between 26 and 32 degrees Celsius, as well as warm nights. Cooling breezes flow across the islands during a dry season that lasts between October and April. The duration of the annual wet season is from May to September, and Palau receives 380 centimetres of rainfall annually. The island chain supports more than 180 different species of birds, in addition to fruit bats, monkeys, and crocodiles, of which both estuarine and salt-water crocodiles are present. The islands are notable because some small saltwater lakes in the interior of one of the islands support jellyfish and other marine organisms that exist in an isolated and unique marine ecosystem.

Only eight of the Palauan islands have a permanent population. But they are all lovely islands with renowned natural beauty, and the waters surrounding the islands are home to many of the world's most colourful sea creatures. In fact, Palau is known throughout the scuba diving community for its fantastic marine life, and scuba divers come from all over the world to dive there. Some 1500 species of tropical fish are native to Palau, in addition to 750 identified species of coral and sea anemones, all of which make Palau a diver's paradise. Offshore, there are blue holes, sea walls, and sheer drop-offs, as well as caves and shipwrecks to investigate, all contained in waters having excellent visibility. The famous French oceanographer, Jacques Cousteau, thought that Palau's reefs were the most complete coral reef ecosystems anywhere on the planet. And the clear visibility in Palauan waters ensures that many different species can be seen by divers and snorkelers. On my flight to Palau, I was pleased to once again see that Jim and Susan, two American diving acquaintances whom I had originally met on Yap, were also travelling to Palau, and I was quick to renew my acquaintance with them. They had heard much about the diving in Palau and they were excited to be able to see it all first-hand.

The Palau Islands are geographically part of the Caroline Islands but they possess a different geological history, and they are ethnolinguistically distinguished from the remainder of the Caroline Islands that lie within the Federated States of Micronesia. As a result, when the future of the Trust Territory was being debated the islands of Palau opted for a separate political status from the other parts of Micronesia. A republican form of government modelled on the United States was created but at the time of my visit in 1989, Palau, unlike the other island chains in Micronesia, had not yet been released from the UN Trust Mandate by the United States. Thus, Palau was the last remaining part of the Trust Territory of the Pacific Islands that was still being actively administered by the United States of America on behalf of the United Nations.

My tour company, Trip-n-tour, had called Palau perhaps "the most beautiful place in the world" and in truth I did find it to be a gorgeous destination. There were beautiful white sand beaches along the coast, and caves and waterfalls on the main islands. In the lagoon off Koror, the capital city lay the famous Rock Islands, mushroom-shaped, and covered with emerald green foliage. Originally, the Rock Islands were part of the reefs surrounding the islands but millions of years ago volcanic activity had forced the reefs above the surface of the Pacific and as the aeons passed vegetation took hold while erosion undercut the base of the Rock Islands where the land met the sea. As a result of the erosion at their waterlines, the Rock Islands appeared to float upon the sea, making the small islands look like giant green puffballs.

The Rock Islands of Palau

The islands seem to float on the ocean like emerald-green puffballs

Palau was settled more than 2000 years ago by waves of migration from the East Indies and Melanesia, with later additions from East Asia and Polynesia. Palau's earliest contact with the west came in the late 1700s, with the arrival of British mariners but the Spanish had laid nominal claim to the islands much earlier. However, it was not until 1885 that Spain established actual colonial administration over the Palau Islands, and then Spain only managed to maintain its control over the islands until 1899, when Spain sold all of its Micronesian possessions to Germany in the aftermath of its defeat by the United States in the Spanish-American War. Although Spain's colonial control had been haphazard at best, Spanish Jesuit priests did bring Christianity to the islands, and they gave the Palauan language an alphabet.

The subsequent German colonial period was marked by a great increase in commercial development. Germany began to establish phosphate mines and coconut plantations, and the Germans made concerted efforts to improve public sanitation in Palau. As with other Pacific island cultures, the early contact between the Palauans and the men of the west had resulted in the introduction of many infectious diseases to which the Palauans had no natural immunity. As a result, the population of Palau dropped from an estimated 40,000 in the nineteenth century to only 4,000 in the early twentieth century. Then, in 1914, upon the outbreak of World War One, Japan occupied Palau and expelled the German colonial presence. Thereafter, Japan retained Palau for the duration of the war, and Japan was eventually given a League of Nations Mandate over the islands at the Paris Peace Conference in 1919. In the interwar years, Japan promoted agriculture and fisheries to enhance the islands' economies but it also took important steps to fortify the islands, causing the displacement of some Palauans from their traditional villages as part of that process. During World War Two the main Palauan islands of Babeldaab, Arakabesang, and Koror were bombed by the United States but they were not invaded. However, two of the outlying islands (Peleliu and Angaur) were invaded by US forces in 1944, and fierce fighting took place, particularly on Peleliu where a fanatical Japanese resistance led to the loss of many lives. Following Japan's surrender in 1945, the United States took possession of all of the Palau Islands. The following year the United Nations conferred a strategic trust upon the United States for all of Micronesia, and Palau largely disappeared from the public eye behind the shield of the US armed forces.

However, as the worldwide decolonisation movement erupted in the 1960s, the US was forced to bring forward the political development of Micronesia towards possible independence. A Congress of Micronesia was established by the Americans in 1965, but as subsequent proposals for autonomy developed it became clear that different parts of Micronesia wished to go their separate ways, among them the Palau Islands. While the Northern Mariana Islands, the location of the headquarters of the Trust Territory, opted to join the United States as a so-called American commonwealth, which is a type of US territorial government, the Marshall, Caroline, and Palau islands all opted for political independence. Due to the artificial economy that the US had created and fostered in Micronesia, as well as a strong desire on the part of the United States not to let the islands remove themselves entirely from the US political orbit, arrangements were made for the three nations proceeding towards independence to enter into Compacts of Free Association with the United States. In return for American military protection, and American economic subsidies, the new Micronesian nations would generally follow the US lead in international relations and protect US interests in Micronesia. While the *quid pro quo* presented no real problems for the Marshall Islanders or the inhabitants of the FSM, it did present a political problem for the Palauans because under the constitution that had been agreed upon by the people of Palau and put into effect in 1981, there was an absolute ban on the use, holding, or storage of nuclear weapons within the territory of Palau. The United States determined that the constitutional provision in Palau prohibiting the presence of nuclear weapons was inconsistent with the proposed terms of the Compact of Free Association between Palau and the United States. Accordingly, the United States demanded that the Constitution of Palau be amended to remove the nuclear weapons prohibition. However, the constitution of Palau could only be amended by a referendum at which 75% of the total votes cast by the electorate in Palau were cast in favour of any proposed constitutional amendment.

In referendum after referendum, the people of Palau failed to approve a measure to remove the nuclear prohibition by the requisite 75% majority, and consequently, the United States refused to grant Palau its independence. Palau thus remained mired in a United Nations trusteeship vested in the United States. After independence was granted in the Eighties by the US to the Marshall Islands and the Federated States of Micronesia, and after the Mariana Islands joined the United States, Palau remained the last UN Trust Territory anywhere in the world. When I arrived in 1989, the ultimate authority over Palau was the US Secretary of the Interior in Washington, and a temporary Office of Transition headed by a Director continued to function out of the Trust Territory offices on Saipan to carry out the duties previously vested in the defunct High Commissioner of the Trust Territory of the Pacific Islands whose office had already been abolished by the United States when the other island groups in Micronesia made their new arrangements with the USA, to the satisfaction of Washington.

In addition to the uncertainty about independence, there was also a significant ferment in domestic politics in Palau. Although traditional village and clan structures continued to have influence the strength of traditional modes of social organization was lessening under the inroads made by the new political structures. New democratic systems were introduced by the United States as part of the process of preparing Palau for independence, and the competition between older traditional political structures and the new democratic politics may have been a factor in the degree of domestic political instability that occurred in Palau in the Eighties. The first President of the new republican government established by the United States under the trusteeship mandate was assassinated in 1985, and the republic's third president committed suicide while he was under a criminal investigation in 1988.

That was the political backdrop when I visited Palau in 1989. Another referendum on constitutional change was then planned. On the drive to my hotel in a taxi cab from the airport, I sensed reluctance on the part of the driver to discuss politics in Palau when I brought up the subject and I thought that he was probably mistaking me for an American. Once I told him that I was a Canadian he opened up with passion about the political stalemate facing his country. He told me that six times previously a majority of the people of Palau had voted to endorse the constitutional amendments sought by the United States in order for Palau to obtain its independence but each time the amendments had not passed because of the constitutional rule that a referendum must be approved by 75% of the voting electorate. In the absence of approval of the changes to the Palauan constitution that it sought the United States refused to grant Palau its independence. My taxi driver clearly felt that his country was being bullied by the United States. But the USA remained adamant. My driver's own personal sentiments were against nuclear weapons, and he talked at length to me about the experiences of Palau in World War II when Americans and Japanese had fought each other over control of these islands, and how many Palauans had suffered despite having no real interests in the conflict.

On the drive from the airport, one sees the many bridges connecting the central islands of Palau

Nevertheless, despite some antipathy towards the US over the independence issue, I found that modern influences had now become paramount in the country. The common language was Palauan, but English was widely spoken (and many older Palauans could also speak Japanese). The US dollar was the local currency here, as it was in the other parts of Micronesia, and the form of government in the constitution proclaimed in 1980 was that of a republican state based on the political structures of the American government. The population in 1989 was only 12,000, guaranteeing that Palau would be viewed as a micro-nation by the international community when it eventually received its independence.

Like the people on Yap Palauans also chew betel nut but they were not as addicted to the substance as the Yapese. The capital town, Koror, was fairly modern and it also boasted some industrial development in the form of a fish cannery. Traditionally, the Palauans built thatched structures that were similar to the structures built in other parts of the Caroline Islands, particularly Yap. Like the Yapese, the Palauans also built village Men's Houses, called an *abai* in Palau, with a design that was similar to those on Yap but a more ornate example, with an elaborate decorative gable typically embellished with traditional Palauan symbolism including references to the legends or customs of Palau. In 1989, when I visited Palau, there was still a prominent traditional social structure in Palau. Each village retained its own hereditary chief and there were two paramount chiefs for the entire island chain. Despite democracy being superimposed on Palauan society, the influence of the chiefs was still considerable.

I found the Palauans to be an ambitious people and at the time that I visited Micronesia the Palauans had produced many more indigenous professionals than the people of other Micronesian islands. Throughout Micronesia, the Palauans were known to be industrious and they had established themselves in commerce and professional capacities on some of the other Micronesian islands. While Palau was certainly not over-developed in 1989, it was definitely at a stage of transition between an idyllic paradise and a modern nation. Unlike other islands in Micronesia, there was no toplessness in dress for either men or women, and both sexes favoured western style casual clothing. Palau was still a conservative society, however, and bathing suits were only regarded as suitable to wear on the beach and they were not considered to be appropriate to wear in the town, or in the villages.

Palauan traditional cuisine included melons, papaya, taro, yams, shellfish such as coconut crab and lobster, and various ocean fishes such as parrot fish, tuna, and wahoo. Traditional Palauan handicrafts included baskets, tortoiseshell jewellery, and woven pandanus hats, purses, and baskets. The most striking Palauan handicrafts were the hand-carved storyboards that provided a portable adaptation of the carved story beams found in the traditional Palauan Men's House, which depict scenes from the legends of Palau. A number of these storyboards were produced by prisoners in the Palauan jail to support themselves during their incarceration.

Upon my arrival in Palau, I found myself at the modern, American-built, air terminal that was located on Babeldaab Island. That island is connected by a bridge to Koror Island, which is the site of Koror, the capital town. Trip-n-tour had arranged for me to stay at the Palau Pacific Resort, a Pan Pacific hotel that was the nicest hotel in Palau. However, upon my arrival at the air terminal, the Trip-n-tour transfer agent advised me, and a couple of other guests, that our reservations at the Palau Pacific resort on Arakabesang island would not be honoured, at least initially, because the hotel was at full capacity. However alternative arrangements had been made for us to stay at the Japanese-owned Hotel Nikko on Koror Island. Upon arrival at the alternative hotel, I also discovered that my luggage was missing except for the small carry-on bag that I had taken with me onto the aeroplane. I checked in at the Nikko, which was a comfortable but standard hotel.

It did, however, offer beautiful views of the marshmallow-like Rock Islands that lay just off-shore from the hotel.

Lodged in an unexpected hotel, and without my luggage left me feeling let down by my tour company but for the moment I had nothing else to do so I wandered out to the front of the hotel where I found that the Hotel Nikko kept two small crocodiles in a pen at the front of the hotel. A simple wire mesh-like fence surrounded the crocodile pen and one of the smaller reptiles was leaning against the fence. Feeling brave but actually being quite foolish I bent down and gently rubbed the jaw of a resting crocodile through the wire fence of its pen. The wire screen seemed to be too small for him to snap at me. Still, it was not the smartest thing to do because I was putting a lot of faith in a wire mesh! Both of the crocodiles seemed very lethargic in the afternoon sunshine, however. Palau is famous for harbouring two types of crocodile, a smaller estuarine crocodile, and the fearsome Australian Saltwater Crocodile, a species in which the males can reach a length of between six and seven metres, and is an aggressive species that are considered to be highly dangerous to humans. Palau, lying at the southwest end of the Caroline Islands is the only part of Micronesia that is within the range of these terrifying reptiles more commonly encountered in northern Australia.

When the Trip-n-tour company representative finally stopped by the Nikko Hotel I complained strenuously to her about the loss of both my hotel booking at the Palau Pacific and the disappearance of my luggage but she maintained that nothing could be done about the dishonoured reservation. However, she did tell me that efforts were being made to find out what had happened to my luggage. She seemed largely unconcerned with my problems but in hindsight perhaps events were beyond her ability to set right, at least for the moment. Later I became a little concerned about remonstrating with her so assertively about my situation when someone pointed out to me that she was actually the daughter of the President of the Republic of Palau! Suddenly, I had a vision of being thrown into a tropical jail to rot in a remote foreign country for arguing with the President's daughter! Eventually, however, my luggage was delivered without explanation to the Hotel Nikko and much later it was explained to me that certain VIP guests from Japan had been staying at the Palau Pacific Hotel and they had decided to stay over an additional day. Since the hotel was full it was necessary to dishonour someone's reservation. I was selected because I was a somewhat younger male travelling alone, and it was thought that I would be quite able to make do elsewhere. However, my reservation at the Palau Pacific Hotel was honoured the next day, and eventually, everything worked out. In the meantime, I settled in at the Hotel Nikko.

The Hotel Nikko, an older hotel, was comfortable if not luxurious. In the hotel pub, I met Carl an American banking assistant vice president based in Japan who had just arrived in Palau for a short vacation with his family. We struck up a conversation and I was interested to learn about his experiences living in Japan as an American expatriate. Carl was just the sort of person that a large US bank might send abroad to gain foreign experience. A tall blond, blue-eyed American with an equally blond, blue-eyed wife and children to complement an MBA from a major US business college. We chatted about his life in Japan and as our meeting occurred around the time of the late Emperor Hirohito's state funeral, we chatted about the Japanese ceremonies attending the death of one Emperor, and the accession of another. The banker mentioned the importance in Japan of the celebration of the Emperor's birthday, which would now have to change from the date of Emperor Hirohito's birthday to that of his son, Emperor Akihito, and which the banker said would be a much less desirable date for a holiday. That led me to note that in Canada we celebrated the official birthday of the monarch on a fixed day in late May, regardless of the monarch's actual birthday, because the May holiday was a convenient marker for the start of summer in Canada. So I suggested that the Japanese might wish to look to Canada as a pragmatic model on the subject of honouring a monarch's birthday.

Soon we were joined by a colourful local character, Francis Torobiong, the proprietor of the local Fish-N-Fins dive shop. Carl wished to make arrangements for a scuba dive in Palau. Francis was a gregarious character and we talked at length about the diving possibilities in Palau. Palau was a great place to scuba dive or to snorkel he said because the reefs were populated with all manner of sea life, and there were even possibilities for wreck diving as a result of the US bombing of Koror on March 30, 1944. On that date, aircraft carriers from United States Naval Task Force 58 had launched a massive attack against Japanese support vessels at Palau with dive bombers and torpedo bombers. The American attack went on for a day and a half and dozens of Japanese ships were sunk. Although some of the sunken ships were later salvaged by the Japanese during the Korean War most of them remain where they were sunk, and they are near enough to the surface for scuba divers to explore them at length.

Our conversation turned to hotels and Carl said that like me he too had wanted to stay at the Palau Pacific Hotel but he had found it to be fully booked when he attempted to reserve a room on short notice. Francis tried to help out Carl, now his new client, when he learned that I would relocate to the Palau Pacific the next day by offering me some free scuba lessons if I gave up my reservation to Carl. I was tempted by the offer but I knew that my stay in Palau would not a long one and that I would probably not get very far towards a diving certification in the time I had in Palau. In the meantime, I did not want to spend all of my time in Palau on a boat so I declined his offer, although later I did question my decision.

Rowdyism was a problem in Koror when I visited in 1989, particularly on Friday nights, and it just so happened that my stay at the Hotel Nikko was on a Friday night. The hotel information booklet given to me at registration stated very clearly that: "The Island Government has found it necessary to impose curfew hours at night on both the local islander and the visitor. Please see the front desk for hours." I learned that Hotel Nikko had a security lockdown at 11 PM. I did venture out in the early evening in Koror and I discovered that mostly there was a lot of noise as young people, both male and female, got together to let off some steam while riding around the streets of Koror in the back of pickup trucks blaring loud music from their boom boxes. To me, it just seemed like youthful high spirits, and I experienced no real danger although I also made sure to be back at the Hotel Nikko before the hotel curfew, so I cannot say what might have gone on in the hours after midnight when their youthful high spirits were probably at their height, after being mixed with alcohol.

The next day I was able to transfer to the Palau Pacific Hotel, a luxurious resort on Arakebesang Island that is connected to Koror Island by another bridge. Arakebesang Island was also the site of the Office of the President of Palau, and therefore the local security was much tighter on that island than in the capital town. The Palau Pacific was located in a beautiful spot where there was a lovely white sand beach and a clear view of the Pacific Ocean. All was forgiven concerning my earlier dishonoured reservation of the previous day when I checked in to my room and found that the resort manager had thoughtfully provided a bottle of champagne and a fruit basket, together with his personal expression of regrets over the hotel's inability to accommodate me on the first day of my stay in Palau.

The Palau Pan Pacific Resort boasted 100 air-conditioned deluxe rooms and suites with private verandas, a beautiful white sand beach, tennis courts, a botanical hiking trail, and a fish pond stocked with tropical fish and turtles. For water sports, the hotel offered catamarans, canoes, snorkels, scuba, and windsurfing equipment so that guests could take full advantage of the pristine waters offshore from the resort. This was definitely a high-end tropical resort and it was a nice place to relax at this point in my journey through the islands of Micronesia. The staff at the Palau Pacific was made up of quiet, well-trained, and helpful young Palauan men and women.

The Palau Pacific Resort

A comfortable garden room at the Palau Pacific Resort

Swimming at the Palau Pacific beach was a delight and I took full advantage of the snorkelling there, where I encountered many species of beautiful tropical fish while snorkelling just off the beach. At the Palau Pacific, I also reconnected with my diving acquaintances from the United States, Jim and Susan, the chemist and his marina manager travelling companion, and I heard from them about the wonders of diving in Palau when, in the evenings, we congregated at the open-air bar at the resort. Other divers also enjoyed an evening out and I heard languages from around the world among the scuba divers. One evening though, as I sat with my diving acquaintances at the open-air bar we were joined by an older couple from the US Midwest. The gentleman was a physician and a veteran of World War Two who had been a prisoner of war in Germany and he still held some harsh memories of those days. I am afraid he had a little too much to drink and he began to take vocal umbrage at the presence of some German divers at a neighbouring table. Fortunately, the Germans at the nearby table ignored his pointed comments, which might have otherwise led to a new war!

Both Jim and Susan had been particularly enthralled by one of their diving experiences when a large school of manta rays had silently 'flown' over them in the ocean. Manta Rays are gentle giants of the sea with wingspans that can reach 3.6 metres and they can weigh up to 680 kilograms. Both Jim and Susan said that it was a magnificent sight seeing a flock pass above them in the ocean and one that absolutely confirmed their decision to dive at Palau. Listening to them describe the sights underwater I began to wonder if I had been too hasty in not taking up Francis Torobiong's earlier offer to give me scuba lessons if I gave up my reservation at the Palau Pacific to his American banker client!

The beauty of the beach on Palau

The garden at the Palau Pacific Resort

Sunset at the beach

I had been quite taken by the beauty of the Rock Islands, which I had seen clearly during my stay at the Hotel Nikko, and I decided to participate in a tour of those islands by boat arranged by Trip-n-tour. When I arrived at the dock in Koror for the tour I found that the President's daughter was in charge of the tour. She took a few of us out on a runabout and we explored the beauty of the Rock Islands up close. On the way out, she stopped the boat in a channel in the Rock Islands and showed us the wreckage of a Japanese Zero warplane that had been shot down by the Americans during World War II, and was now resting in the shallow clear waters of the Rock Islands where it was clearly visible from our boat.

The turquoise beauty of the waters surrounding the Rock Islands

A close-up photograph of the Rock Islands shows the effects of the erosion at their base

Touring the Rock Islands by boat, 1989

On another day I visited the Belau National Museum at end of Hospital Road in Koror. The museum showcased Palau through historical artifacts and exhibits emphasizing Palauan history and culture. A handicraft shop attached to the museum sold books with a Palauan theme, as well as leis, coral and shell jewellery, T-shirts with local design motifs, and watercolour paintings. A particular feature of the store was the carved wooden storyboards that Palau was famous for. These storyboards consisted of a wooden plank that displayed carved figures depicting scenes from the history and legends of Palau. I saw such storyboards elsewhere in Micronesia and I purchased one in Yap during my visit to that island but the Palau storyboards were much more elaborate than the storyboards on Yap, and they were also carved from heavier tropical woods. Although Palauan storyboards are a small-scale example of the elaborate artwork traditionally installed on an *abai*, the storyboards were only popularized during the period of Japanese rule when a visiting Japanese artist came to the islands and encouraged the local indigenous population to begin carving them. They now constitute a unique form of folk art and I was soon interested in acquiring one.

According to the store manager at the museum, the storyboard inventory in the museum store was all carved by the prisoners in the local Palau Prison, which I had observed earlier during one of my walks about Koror Town. I was told that the prisoners had to support themselves while they were in jail, and the artistic storyboard carvings for locals and tourists was a socially acceptable way for them to do that. At the museum, I bought one of the elaborately carved storyboards carved on a heavy dark wood by a prisoner, which depicted one of the many legends of Palau, the legend of Remerang and the Morning Bird.

A Palauan storyboard depicting the legend of Remerang and the Morning Bird

A detail from a Palauan storyboard showing an abai, or Palauan Men's House

Detail from a Palauan storyboard depicting women interacting

Life in Koror seemed quite casual though I suppose it was not for its inhabitants and perhaps it only seemed that way to me as an outsider because the pace of life was far more relaxed on these islands than in North America. One day in Koror I decided to rest under a coconut tree. A young Palauan man was sitting on the other side of the tree and he was dressed almost as casually as I was. I struck up a conversation with him and to my surprise, it turned out that he was the chief of protocol in the foreign office of the Republic of Palau who was taking a short break from his duties. After learning that I worked for the Canadian government's aeronautical authority he offered to set up meetings for me with Palauan officials since Palau was anxious to increase international air services to the islands. I had to demur because I had no authority to discuss air rights with foreign officials while engaging in personal travel and doing so would definitely have caused difficulties for me back home. Still, I marvelled at the casualness of life in the capital of Palau, and how one could easily make the acquaintance of Palauan officials while sitting under a coconut tree!

On another day I went out on a boat for a snorkelling expedition off the reef that encircles Palau. Again, the President's daughter was in charge. On the reef, we saw more of the tropical fish life that Palau was famous for but the high point of our expedition was seeing Giant Clams directly below us on the reef. There are several species of Giant Clam present in the waters of Palau. We could clearly see their colourful mantles and it was fascinating to see these largest of all mollusks in their natural habitat. Giant Clams can grow to a width of 1.2 metres and can weigh up to 90 kilograms and they can live up to sixty years. The President's daughter carefully warned everyone not to touch or disturb this unique and protected form of marine life in any way.

Traditional structures., Palau

If Yap had been fascinating for its vibrant traditional culture my trip to Palau was memorable for the physical beauty of the Rock Islands, and the fascinating ocean creatures that I saw while snorkelling. The Palauans had readily assimilated modernity, in the form presented to them by the United States as the UN Trustee power but they had not wholly abandoned their island traditions. Nevertheless, Palau was an island culture in transition and Palau was more than eager to accept the fruits of a modern style of life.

My next trip would take me to the Micronesian north, to the Commonwealth of the Northern Mariana Islands, and to the island of Saipan, an island of bitter memories.

SAIPAN – THE ISLAND OF BITTER MEMORY

Saipan is the major island of the Commonwealth of the Northern Mariana Islands ("CNMI"), a US possession that emerged from the United States Trust Territory of the Pacific Islands. The Northern Mariana Islands consist of the tips of a massive underwater mountain range that rises about ten kilometres from the Pacific seabed. All of the Northern Mariana Islands are volcanic islands, with some of their volcanos still considered to be active. The islands stretch across 650 kilometres of the North Pacific, north of the island of Guam, with the two most northerly islands actually lying above the Tropic of Cancer, placing them in the temperate climatic zone instead of the tropic zone. Although the island of Guam is geographically part of the Mariana Islands, it is politically separate from the CNMI, and therefore the island of Saipan has become the capital island of the CNMI.

The southernmost islands of Saipan, Tinian, and Rota are the largest islands in the CNMI and the only islands that can boast a substantial population. They also contain some of the best beaches in all of Micronesia. The Northern Mariana Islands are much drier than other parts of Micronesia, and one consequence of the drier climate is that the striking Flame Tree grows here in profusion. Alas, when I visited Saipan it was not yet the season for their leaves to turn crimson.

Saipan claims that it possesses the evenest climate in the world. Temperatures on Saipan hover consistently between 28 and 29 degrees Celsius year-round. Saipan is also the sunniest island in all of Micronesia and rains there are infrequent, mostly occurring between July and October of each year. However, the Mariana Islands are also located on a well-established Pacific typhoon track, and occasional typhoons do make landfall on Saipan. In part, perhaps, because of the seasonal typhoons, there are quite a few ancient shipwrecks around Saipan because the island was an important stopping port on the east-west navigation route followed by the Spanish treasure galleons voyaging between Peru and the Philippines in the seventeenth and eighteenth centuries. It is quite possible that there may still be Spanish treasure ships lying undiscovered in the waters of the Northern Mariana Islands.

Saipan was originally settled by the Chamorro people between 1500 and 2000 BC. Western first contact came in 1521 when Ferdinand Magellan visited Saipan on his historic expedition's first circumnavigation of the Earth. Subsequently, Spain concentrated its colonization efforts on the Mariana Islands and the Spanish succeeded in establishing a firm colonial control over Saipan. Spain sent Jesuit missionaries to the Mariana Islands and the Jesuits brought western culture and western moral norms to the islands. When opposition to Spanish rule inevitably surfaced on Saipan the entire population was forcibly displaced to Guam, and the Spanish replaced the Chamorrans on Saipan with Carolinian islanders. Later, many of the displaced Chamorrans returned to Saipan and engaged in farming, while the now-resident Carolinians engaged in fishing, allowing both groups to get along with each other relatively harmoniously.

Spanish colonial rule continued uninterrupted throughout the entire Mariana Islands until 1898 when the United States seized the southernmost island of Guam during the Spanish-American War. The following year, after its defeat by the Americans, Spain sold the remainder of its Micronesian possessions to Germany. Upon the outbreak of World War One in 1914, the German naval commander in the Pacific, Vice Admiral Count Maximilian von Spee assembled his far-flung East Asiatic Squadron at Germany's remote auxiliary anchorage at Pagan Island in the Mariana chain after Japan's entry into the war on the side of the allies prevented him from using his primary naval base at Tsingtao in China. At Pagan, Admiral von Spee decided to dash across the Pacific, round Cape Horn and play hide and seek with the British Royal Navy in the Atlantic in hopes of reaching Germany. Von Spee's powerful armoured cruisers *SMS Scharnhorst*, and *SMS Gneisenau*, together with his light cruisers overpowered and defeated a British navy squadron looking for him at the Battle of Coronel off the coast of Chile, giving Germany its most significant naval victory in the war. Von Spee's victory enraged the British who then sent two giant battlecruisers to the South Atlantic that tracked down von Spee after the Germans rounded Cape Horn and entered the Atlantic and destroyed von Spee and his squadron in the Battle of the Falkland Islands.

Meanwhile, the Germans were quickly displaced by the Japanese from the Northern Mariana Islands, and the Japanese settlement of the Mariana chain soon followed. During the interwar years, the Japanese put Saipan under intensive agricultural cultivation. Sugar cane became the main agricultural crop, and sugar refining the main industry under Japanese rule. The Japanese restricted foreign access to their South Seas Mandate, the entirety of the Micronesian islands

Japan held under the authority of the League of Nations. However, the United States remained suspicious of Japanese intentions and there was some limited American intelligence gathering within the Japanese mandate during the interwar years. Such efforts may have been the basis for an old and intriguing rumour concerning the Japanese jail on Saipan. There were rumours that the Saipan jail had once held Amelia Earhart, the famed American aviatrix who went missing over the Pacific Ocean on July 2, 1937, while she was making an ill-fated attempt to fly around the world with her navigator, Fred Noonan. In the years since some conspiracy buffs have alleged that Earhart was on an intelligence-gathering mission for the US government and that she was shot down and captured by the Japanese, before being incarcerated on Saipan. However, there is no firm historical evidence about the fate of the famous aviatrix, and no reliable historical evidence that she was ever in Saipan, so the rumours of her presence remain mere conjecture.

In 1941, upon the outbreak of World War Two in the Pacific Ocean Japan conquered Guam and thus the entire Mariana Islands chain once again came under the sovereignty of a single colonial power. As the war progressed the United States Navy made plans to invade the Mariana Islands to use them as bomber bases from which to bomb Japan. In 1944, US forces invaded Micronesia, taking Kwajalein, Majuro, Peleliu, Saipan, Tinian, and Guam in bitter battles. The loss of Saipan was an especially bitter blow to the Japanese and after the fall of Saipan, Japan began to face the prospect of defeat at the hands of the allies.

Tinian Island is only five kilometres from Saipan and it is clearly visible from Saipan. During World War Two it was chosen as the main bomber base for the US Army Air Force. American naval Seabees quickly constructed four 2.5-kilometre runways to accommodate the giant US Army Air Force B-29 Superfortress bombers that launched devastating raids against the Japanese home islands. It was from Tinian that the atomic bomb attacks on Hiroshima and Nagasaki were launched in August, 1945. Shortly thereafter, the Japanese Emperor, Hirohito, ordered the Japanese armed forces to surrender to the allies, and the war ended on September 2, 1945, when US Army General Douglas MacArthur accepted the Japanese surrender aboard the *USS Missouri* in Tokyo Bay on behalf of the allied powers.

A wrecked American tank sits on the reef at Saipan where it was stopped in the 1944 battle

After the war, the United States retained control of the Northern Mariana Islands as part of the strategic trust over Micronesia granted to the USA by the United Nations. From 1945 until 1962 the Northern Mariana Islands were under the jurisdiction of the US Navy. After 1962, the administration of these islands was transferred to the UN Trust Territory administration, which established its headquarters on Saipan Island. When I visited Saipan in 1989, it was still the capital island of the Trust Territory, although by then the authority of the Trust Territory only extended to the far-off Palau Islands.

As the Trust Territory was being wound up in the late Seventies and early Eighties the population of the Northern Mariana Islands decided not to join in a new Micronesian state, nor did the islanders seek to become an independent state in their own right. Rather, the Mariana Islands wanted to remain a part of the United States and the people chose to become the Commonwealth of the Northern Mariana Islands, obtaining an American self-governing territorial status that was similar to the status of the island of Puerto Rico in the Caribbean Sea. A local woman that ran the car rental booth at the airport where I rented a car upon my arrival in Saipan explained the new status of the CNMI to me. She said that in 1987 the CNMI was formally established, and the people of the Northern Mariana Islands received full US citizenship, and a form of government that was modelled on that of the United States with their own elected governor, and an elected legislature, for the islands. American citizens living in the CNMI are represented in the US Congress by a non-voting member of that body. As with other overseas territories of the United States, CNMI citizens cannot vote in US presidential elections because only the citizens of one of the states of the American union are eligible to send delegates to the US Electoral College, and therefore only American citizens resident in a state are eligible to vote in presidential elections.

In 1989, when I visited Saipan, the population of the CNMI was about 17,000, and Saipan was highly developed – a sort of miniature Guam with asphalt roads, cement block homes, modern hotels, radio and television, American banks, ubiquitous automobiles and trucks, and daily air service. Demographically, in 1989, the Mariana Islanders were a mix of Chamorro, Carolinian, Filipino, and Japanese ethnicities. I found the CNMI to be a modern place with western-style hotels that included popular recreational facilities such as windsurfing and sailing, as well as nightclubs. English was the official language and was also widely spoken, although the Chamorro and Carolinian languages were also spoken by residents, and some of the older residents could speak Japanese, which was also spoken by many people employed at the major hotels. As in Guam, the Spanish influence has resulted in a large Roman Catholic population on the island. Dress and deportment of the inhabitants largely followed US mainland customs and norms, and, as in Guam, the local greeting was *Hafa Adai* (literally 'What's Up') to which the locals might respond by subtly raising their eyebrows, or smiling back at you.

The well-developed tourist facilities on Saipan included modern hotels with nightclubs, tropical botanical gardens, and restaurants that catered to multi-ethnic cuisines such as Japanese sashimi, Korean, Chinese, Filipino, Mexican, and the ubiquitous American fast food joints. By far the greatest number of tourists visiting Saipan came from Japan. Saipan was also well-known for hosting annual regattas for both windsurfing and sailing and, regrettably, cockfighting was still legal there at the time of my 1989 visit.

A beach on Saipan, on a day that belies Saipan's claim to be the sunniest part of Micronesia!

I drove along the North-South coastal road of Saipan which is adjacent to the Philippine Sea, and I stopped along the way at the American invasion beaches where US marines had landed under intense Japanese fire in 1944. There was an old American tank lying out on the reef that had been stopped by Japanese shelling during the battle for Saipan in 1944. Now it was rusting relic of the battle that had been fought here. Farther north was the Last Command Post, where the Japanese commanders had directed their final defence of the island during the late stages of the battle. The fighting on Saipan was especially bitter, even by the Pacific war theatre standards, and casualties were very high among the victorious Americans while the Japanese defenders suffered overwhelming losses. The Last Command Post remained a shattered structure when I visited it. I met some tourists from Japan there, and we talked briefly about what had happened here long ago. The Japanese tourists, like me, were taken aback by the level of destruction – the bombardment must have truly been a hell for those on the receiving end of the US navy shelling. But there was hope in our shared horror at the level of destruction. Oriental and western countries, once enemies are now friends, and historic sites like the Last Command Post promote our collective hopes for peace to prevail in the years ahead.

Visitors from Japan tour the Last Command Post, where the Imperial Japanese Army conducted the final defence of the island during the American invasion in 1944

As the American forces closed in on the final Japanese defenders in 1944, the senior Japanese commander, Vice Admiral Chuichi Nagumo, the admiral who had led the Japanese Navy to victory over the United States at the Battle of Pearl Harbour, and over the British at the Battle of Ceylon, before suffering a crushing defeat at the hands of the US Navy at the Battle of Midway, found himself beached and forgotten on Saipan. As his defences crumbled around him Nagumo chose to commit suicide to avoid the ignominy of defeat, as did the senior Japanese

Army general on the island. But theirs were not the only suicides that occurred as the battle for Saipan concluded.

Shell damage to the Japanese Army's Last Command Post from the US Navy's bombardment in 1944

At the north end of Saipan are two great cliffs, one facing outwards to the sea and the other facing inwards across the island. As the Battle of Saipan reached its conclusion large numbers of the Japanese civilian population sought sanctuary in the north where, together with the Imperial Japanese Army troops, they faced a terrible choice. Japanese propaganda had convinced the Japanese troops, and the Japanese civilian population on Saipan, that they would be treated brutally by the Americans. Japanese propagandists suggested that Japanese women would be raped by US marines and soldiers. Many families made a decision to commit suicide rather than face dishonour and then live under a US occupation. At both locations, Suicide Cliff, the name afterwards given to the landward-facing cliff in North Saipan, and Banzai Cliff, which faces seaward, whole families committed suicide, as did many of the remaining Japanese Army defenders.

At Suicide Cliff, which faced landward, and rose to almost 250 metres in height, whole families lined up single file perpendicular to the cliff and placed their youngest children closest to the cliff edge followed by the next youngest and so on until the line reached the oldest child, who was then followed by their mother and then by their father who positioned himself at the end of the line. Once the line was established the second oldest in line pushed the youngest sibling over the cliff and then the next oldest pushed the second youngest over and that went on until all the children had fallen to their deaths, the mother, at last, pushing her oldest child over. Then the father pushed his wife over the cliff and finally, the father would turn around and run backwards over the cliff so that he would not see his last step. As an even more violent alternative, some families gathered themselves together in a circle around a hand grenade and pulled the pin, blowing themselves to kingdom come.

Suicide Cliff on Saipan Island

At Banzai Cliff, which overlooked the Philippine Sea, the same thing happened as Japanese families gathered there. However, unlike the mass suicides at the landward-facing Suicide Cliff, the sailors on the offshore US Navy armada saw what was happening at Banzai Cliff and used their ships' loudspeakers to plead with the Japanese civilians and troops alike not to kill themselves and their families. But it was all to no avail, and hundreds of Japanese civilians and troops were pushed or jumped to their deaths while the sailors on board the US warships watched in horror. It is said that some Japanese parents tied heavy stones to the shoes of their children so that they would not try to swim in the unlikely event that they survived their fall into the ocean. A few, a very few people actually, survived their fall into the ocean at Banzai Cliff and were picked out of the water by US Navy boats patrolling near the cliffs but no one survived the fall at the landward-facing Suicide Cliff. All of those who died were the victims of Japanese wartime propaganda. But Saipan marked the beginning of the end for the fascists that ruled Japan, and after the Japanese defeat in the Battle of Saipan the Premier of Japan, General Hideki Tojo was forced to resign. Although Japan was not yet ready to surrender those who came to power after Tojo were under no illusions about the possibility of a Japanese victory, and defeat began to loom over Japan.

Banzai Cliff on Saipan Island

After the war Banzai Cliff became famous for another reason. Rather than remove war materiel and send it home to America, the US armed forces chose to haul tonnes of war materiel up to Banzai Cliff and then toss it over the side into the ocean, including tonnes of ammunition.

Apparently, it was thought to be too expensive to send it all back to the United States, or so it was said.

A Peace Memorial Park now graces the cliffs where once Japanese, including children, jumped or fell to their deaths. The cliffs are now quiet, ghostly places, where the only sounds are of the wind, and of the tinkling of Buddhist memorials. It is a sad and mournful monument to the horrors of war, and a sombre place of remembrance.

The Memorial Park at the northern cliffs of Saipan where masses of Japanese committed suicide in the final stages of the Battle of Saipan

As I retraced my path on the coastal road, now heading south, I stopped at another invasion beach and I spoke to an older American couple who had stopped to survey the scene, which included, off in the distance, the island of Tinian. The gentleman was a US Army Air Force veteran, one of those who have been described as 'the greatest generation' who fought for freedom in World War Two. As a young man during that war, he had been a B-29 Superfortress pilot and the day before I met him he had been back on Tinian Island, where he had once piloted bombers into the air for strikes against Japan in the later stages of the war. This was his first trip back to the islands since the war ended and I could tell that he was very moved by the memories of those days. He told me that being on these islands brought back many sorrows from that time but he expressed the hope, and one that I think we all share, that such a war might never happen again.

Before ending my all too brief visit to Saipan, I chose to visit one more place on the island, The Grotto. There is on Saipan a beautiful sunken pool connected to the sea by two underwater passages where the water possesses a beautiful blue colour that is reminiscent of blue ink, depending on the light. It is a place where it is possible for scuba divers to enjoy views of colourful tropical fish amidst lace corals. When I visited The Grotto a Japanese scuba diving group had just emerged from their dive and their expressions showed their delight in exploring this marvellous underwater world that Saipan offered.

Japanese scuba divers emerge from an underwater tour of The Grotto on Saipan

I concluded my visit to Saipan with some hope for the human condition from the knowledge that people who were once enemies here now visit this island as friends. I departed Saipan for Guam, the heart of America in Micronesia.

12

GUAM - AMERICA IN THE WESTERN PACIFIC

The island of Guam is situated in the western Pacific about 1450 kilometres above the equator at 13 degrees north latitude and 144 degrees east longitude. It is the southernmost and largest island of the Mariana Islands chain. In size, it is about 48 kilometres in length and ranges between 6 and 14 kilometres in width. Guam has an area of 550 square kilometres and, somewhat uniquely; its origin towards the southern end of the island is volcanic and quite hilly while the north end of the island consists of an elevated limestone plateau with steep cliffs and a narrow coastal shelf. It is sometimes called the 'crossroads of the Pacific' because of its air connections to the Orient, the South Pacific, and the United States. It is also the largest landmass between Hawaii and the Philippines and it offers a duty-free port and relaxing resorts on Tumon Bay with sunshine, water sports, and other activities. But Guam is also an important US military base in the western Pacific Ocean.

The climate of Guam is tropical, with a dry season between December and April and a wet season between July and October. Guam receives up to 215 centimetres of rain annually, and its temperatures range between 21 and 32 degrees Celsius. May and June are the warmest months while December is the coolest. Humidity on the island is tempered by the Northeast Trade Winds but the island is also within the North Pacific typhoon track and therefore typhoons are a perennial threat. It is a lush green island, graced by many flowers including Hibiscus, Plumeria, Bougainvillea (the official flower), and Cup-of-Gold, among others.

Guam is an unincorporated Territory of the United States, which means that it is part of the United States but is external to it, as only the so-called incorporated territories are marked to enter the American union as a state. Supposedly, it is famous as the place where America's day begins because the time zone for Guam is fifteen hours ahead of the Eastern Time Zone on the continental United States, thus making Guam the first American community to meet each new day.

Demographically, the population of Guam in 1989, at the time of my visit, was 126,434 including 23,600 US military personnel. The breakdown of the local population was Chamorro 42%, Caucasian 24%, Filipino 21%, and a smaller scattering of Hawaiians, Filipinos, Chinese, Japanese, Koreans, and Micronesians. The major languages were English and Chamorro with both languages being taught in the schools and used in official documents. However, English was very clearly the more dominant language. Nevertheless, the standard greeting was Chamorro; *Hafa Adai*, (pronounced huh-fah-day), literally 'What's Up?'.

Guamanians are US citizens and they send a non-voting representative to the US Congress as well as electing their own territorial governor but they lack the right to vote for candidates for the US presidency because Guam is outside of the union. It was only in 1970 that the people of Guam obtained the right to vote for their own governor, although previous to that they had been granted an elected unicameral legislature. Nevertheless, despite voting for their own local rulers, the island's government remains subject to the overall supervision of the US Department of the Interior.

The Chamorro people likely came to Guam from what is now Indonesia and Malaysia around 2000 BC. They prospered on Guam, living under district chiefs prior to colonization. There were perhaps 100,000 Chamorros in total when the Portuguese explorer Ferdinand Magellan, sailing for Spain, arrived at Umatac Bay in southern Guam in 1521. Magellan found the Chamorro people to be "tall and robust, not too dark with skin pigmentation somewhere between that of the American Indian and the Oriental. The men were handsome and powerfully built and the women were usually opulent, busted, and graceful."

Guam became a desirable way-station for the Spanish galleons laden with silver from South America on their way to the Spanish Philippines. Given its importance to trans-Pacific navigation, Spain asserted sovereignty over the island in 1565, but Spain did not take physical possession of the island until 1688 when Jesuit priests and a Spanish garrison arrived. Soon after, the desire of the Spanish to impose their own cultural norms and religious values led to a conflict with the local population and a rebellion broke out, causing Spain to reinforce the garrison. The rebellions were not fully suppressed until 1695, and by then the Chamorro males had been decimated, with imported diseases killing off most of the remaining population. It is thought that the Chamorro population fell from 100,000 at the initial point of contact to about 5000 at the end of the seventeenth century. Thereafter, the Spanish garrison and men from the Philippines intermarried with the remaining Chamorro females and the island was repopulated.

Nevertheless, the matrilineal customs of the Chamorro people ensured that Chamorro cultural traditions were passed down through succeeding generations, and the Chamorro culture of Guam was not assimilated into either the Spanish or Philippine cultures. However, Spanish culture did infiltrate Chamorro customs as colonialism became more entrenched. Spain established schools and gave the Chamorros a written language as well as provided the local population with new techniques of agriculture, and construction, but Guam nevertheless remained a colonial backwater in the Spanish Empire.

In 1898, the United States burst onto the world stage as an expansionist power in the Spanish-American War, which resulted in the United States annexing much of the remaining Spanish Empire. Although the United States did not seize the Caroline Islands, or the more northerly Mariana Islands, from Spain it did take the Philippine Islands, and the island of Guam, which was subsequently ruled by the US Navy until World War Two.

In 1941, World War II broke out in the Pacific Ocean. The Guam garrison consisted of 153 US marines, an 80-man local Guam Guard, and 271 US Navy sailors and volunteer naval militia personnel. Japan bombed Guam on December 8th and invaded the island on December 10th with 5000 men and Guam fell almost immediately to the Japanese. Under Japanese rule, the Japanese yen replaced the American dollar, and all cars, radios, and cameras were confiscated. A Japanese curriculum was instituted in the schools and Japanese movies and social events were offered to the local populace. Later in the war when the American island-hopping strategy threatened Japanese control in 1944, all of the schools were closed and the Japanese adopted a stricter and harsher rule over the island, using the local population as forced labour to prepare the Japanese defences. After the defences were completed the majority of the population was confined in camps in the interior, which probably saved many of their lives when the US armed forces invaded Guam in July, 1944. After a sharp campaign, the island was captured by the United States in August. The intense battle for control of the island cost 7000 US casualties, and 17,500 Japanese casualties (out of a total complement of 18,500 in the Japanese garrison). The following year Japan surrendered after the atomic bombing of Hiroshima and Nagasaki, and US rule over Guam was permanently restored.

When I visited the island in 1989, the memories of the war were receding but they were still very much present in the population. I toured the War of the Pacific National Park, which commemorates the American campaigns in the Pacific war theatre during World War II where a young Chamorro man explained to me that Liberation Day remained the most important patriotic holiday for all Guamanians, who recall with gratitude the US effort to wrest back control of their island from the Japanese. In 1989, the national park was still under development but I visited the beach battle sites at Asan Point where the US Marines had come ashore on their quest to liberate Guam. The national park also offered a very good audio-visual presentation about the Pacific War.

The invasion beach at the War in the Pacific US National Park with Japanese coastal gun emplacement

While in Guam I stayed at the Fujita Tumon Bay Hotel on Tumon Bay, an older but comfortable low-level three-story hotel that my travel agent had found for me, persuading me that it offered reasonable rates with beach access. The Fujita was well-positioned on attractive grounds erupting with flowers. The grounds also contained some old coastal artillery emplacements and guns that were left over from the war. Large frescos on the walls of one of the buildings portrayed famous incidents from the legends and history of Guam.

The grounds of the Fujita Tumon Bay Hotel on Guam

A Japanese coastal gun emplacement on the grounds of the Fujita Tumon Bay Hotel

A giant fresco on the wall of the Fujita Tumon Bay Hotel depicts the legends of Guam

At the Fujita, there was a fine beach with soft white sand and warm waters on beautiful Tumon Bay, although the beach was somewhat narrow at high tide. Fujita Beach boasted calm waters and a lack of crowds, which was exactly what I looked for in a beach resort. There were other higher-end resorts located on Tumon Bay but they were located farther away, giving the Fujita beach some measure of privacy. In fact, there were relatively few people at the beach during my stay despite the fact that the weather was lovely and I sometimes had the beach to myself. I attributed the lack of beachgoers to the fact that almost all of the guests staying at the Fujita were from Japan, and the Japanese tourists seemed to be uninterested in the use of the beach, although some, particularly children, did use the decent-sized outdoor pool that was located on the hotel grounds.

The Fujita Tumon Bay Beach, showing Puntan Dos Amantes (Two Lovers Point) in the background

At the hotel, there was a large restaurant where I felt somewhat like a celebrity as I ate in splendid isolation surrounded by Japanese families. Many of the small Japanese children stared at me as the sight of a Caucasian person was no doubt an unusual experience for them. The number of tourists from Japan was not surprising because the tourist trade in Guam largely caters to Japanese tourists. In 1987, two years before I visited, tourist arrivals reached 484,000 and 85% of all arrivals came from Japan.

At the hotel restaurant, I met a young Micronesian waitress who hailed from Pohnpei and who was surprised to learn that I had recently been to Pohnpei, and even more surprised to learn that I had actually visited Nan Madol despite the ancient superstitions surrounding the site. The ancient ruins of Nan Madol are often avoided by the locals of Pohnpei, who believe that mysterious dangers can befall those who visit it. Although my waitress catered to the largely Japanese tourist trade in the hotel restaurant she expressed a dislike for them because they did not leave tips! She demonstrated ambition, however, and the fact that she had left her home island to come to Guam to seek better economic opportunities attested to that. Although she worked as a waitress at the hotel restaurant during the day she also sang in a band at a local pub several nights a week. But she missed Pohnpei, especially now that she only returned to her home island infrequently.

Sunset on Tumon Bay, Guam

The fare at the Fujita restaurant was either standard Japanese or standard American fare, and there were no local Chamorro options. For that, I had to rent a car, a nice blue Ford Mustang convertible, and travel south on Marine Road, which is the main circular road in Guam. South of Agana, the capital city of Guam, I found a local restaurant that had been recommended to me by a woman I met earlier on Yap for Chamorro-style food. Chamorro cuisine involved dishes such as barbecued spare ribs, and chicken spiced with finadene sauce and served with coconuts, rice, and taro. The key to Chamorro cuisine, according to the proprietors of the restaurant I visited, is hot finadene sauce. Finadene sauce, a spicy condiment, is the signature ingredient of Chamorro-style food and is made from a mixture of lemon or vinegar juice, soy sauce, onions, and chilli peppers (and occasionally tomato), though there are also other recipes. Chamorro cuisine has been heavily influenced by Spanish, Filipino, and even Japanese cuisines, and it is perhaps now more of a fusion of Chamorro and other styles. I learned later that the best examples of Chamorro-style food were to be found in the public market, rather than in restaurants because both the tourist industry, and the American military presence on the island, have demanded various American, Japanese, Chinese, Filipino, or even Mexican cuisines.

With my rental car, I was able to see the main points of interest on the island, starting with Agana, the capital, which is now called Hagatna. Agana was a pleasant town on the west coast of the island a short drive south of the Tumon Bay resort area. The Spanish influence in Guam was still very present and the prominent Roman Catholic Cathedral gave clear evidence of that. A few years before I visited Guam the peripatetic Pope John-Paul II had visited Guam, and he had spoken publicly to a crowd of several thousand people packed in front of the cathedral.

Agana (now Hagatna) is the capital of the island of Guam

Nearby, and dominated by the cathedral, was the historic Plaza de Espana, which at one time was the centre of Spanish colonial rule. Although heavily damaged during the war, several structures remained within the park-like plaza and I visited the charming Chocolate House, where the wives of the Spanish governors had once served afternoon cocoa to their guests. All that remains of the Spanish governors' official residence now is the Azotea, a raised terrace, and the Garden House, which I found to house a small museum of Guam's history, including displays about a famous Japanese straggler from WWII, Shoichi Yokoi. I remembered reading about this man when I was a teenager. Apparently, Shoichi Yokoi, a Japanese soldier acting out of extreme loyalty to Emperor Hirohito had hidden in the interior of Guam for decades after the war and only presented himself to the Americans to surrender in the Seventies, almost thirty years after the formal end of World War Two. The Plaza also contained a small gazebo dating from the Spanish period called the Kiosko. Three stone arches are all that remained of the former Spanish arsenal. Surrounding the Plaza de Espana were stone walls dating from the Spanish period, enclosing the pleasant park.

The Plaza de Espana in Agana, Guam

The Chocolate House in the Plaza where the wives of Spanish Governors entertained

Near the Plaza de Espana is Latte Park. The lattes are ancient stone pillars that may have been once used as foundational supports for Chamorro houses, and meeting places. However, the early explorers who reached Guam never found any contemporary houses built upon them, so it is also possible that the pillars served a religious, or cultural, purpose. Today they are all that remain of the structures built by the earliest inhabitants of the Mariana Island chain. The latte stones at Latte Park were brought to Agana from the southern interior of Guam and a collection of them stands as mysterious sentinels in the park.

The Latte Stones in Latte Park, Agana, Guam

A short drive inland and upwards in altitude brought me to Government House. Built in 1952, it was the official residence of the Governor of Guam and I was given permission by the house staff to examine the small museum room within Government House. The small collection of objects at Government House contained some interesting carvings as well as the official gifts given to the Governors of the territory. While I was browsing in the collections room I could overhear the First Lady of Guam (i.e., the Governor's wife) speaking in an adjacent room to the local Guamanian contestants in the Miss Universe beauty contest. The Miss Universe pageant was important in Guam and the First Lady was counselling the local beauties on how to make a good impression throughout the selection process both on behalf of themselves, and on behalf of Guam.

A carved sailing canoe in the small museum room inside Government House, Agana, Guam

I left Government House and drove back to Agana to connect with the Marine Road and then I drove south, passing the US naval station before I emerged into a hilly country surmounted by winding roads where the early Spanish influence was still very pronounced. I found a good example of the Spanish influence at Umatac, the pleasant village where Ferdinand Magellan first made landfall on Guam in 1521. Arriving in Umatac I found myself driving under striking Spanish-style arches that one might see in the countryside in Spain, commemorating Magellan's arrival. While most of the north and central part of the island is the preserve of major military installations, the south is given over to agriculture and I discovered that one of the more popular crops grown on the island was watermelon, mostly grown on small farms.

The villages in Guam often have weekend celebrations focused around Roman Catholic saints and every village has its own patron saint. Saint Day fiestas are both a religious and a social event and they are invariably held on weekends, with Sunday being set aside as the feast day, and the preceding Saturday devoted to preparations. A formal procession through the village is generally held on Saturday and the local Roman Catholic church will celebrate a mass commemorating the saint on the next day. These patron saint celebration weekends provide an opportunity for an entire village to engage in socializing with their neighbours, and thus help to strengthen community bonds.

Hilly country in southern Guam

The Spanish-style Bridge at Umatac in southern Guam recalls the arrival of Ferdinand Magellan in 1521

The Chamorro people that I encountered were friendly, if perhaps slightly reserved. One charming cultural difference I discovered when dancing with Chamorro women was their custom of dancing in a large circle around their male dance partner. In 1989, when I visited Guam two of the local pastimes were cock-fighting and Greyhound dog racing, neither of which I attended. Cockfighting was a locally popular but vicious activity. A cock would be trained for up two years, receiving vitamin supplements and hormonal treatments to build up its resilience and given access to hens in order to strengthen its aggressive tendencies towards other males. According to the local view, a good fighting cock was one with a solid build, erect profile, long wings and thick feathers. When chosen to fight they would be equipped with a small knife attached to their left leg and the cocks would be placed in a small pen to battle with their challenger, while spectators gambled on the outcome. This was a blood sport, and often there was only one survivor, and occasionally none. Although widely practised at the time of my 1989 visit this indefensible sport has since been banned altogether. Greyhound racing was also available in 1989, at Guam Greyhound Park and it attracted locals as well as some visitors who gambled on the racing outcomes. In the years since my visit attitudes to dog racing have changed considerably for the better and dog racing in Guam is now also banned.

After reaching the southernmost part of Guam on my driving trip around the island I began my return via the east coast of Guam and I stopped at Inarajan (now Inalåhan) a historic village on the south-east coast that has preserved much of its Spanish heritage. At Inarajan I viewed the natural pool that had formed at the seaside, which had been improved so that people could enjoy a calm saltwater pool with diving platforms and arched bridges that were protected from the fury of the ocean waves by lava rock. The park had been improved by seawalls and the little stone bridges as well as by the installation of showers, making this a desirable place for a visit by the many families I saw enjoying a day at the seashore. Nearby is Mount Lamlam, which some suggest is the tallest mountain on earth if one measures its height from the bottom of the Mariana Trench, the deep marine trench that runs parallel to Guam in the North Pacific Ocean, and which is the lowest part of the Earth. However, I think that Hawaiians might challenge the Guamanian's description of Mount Lamlam as the tallest mountain because Hawaii touts Mauna Kea, a mountain on Hawaii Island, as the tallest mountain on Earth as measured from the level sea floor. Of course, Mount Everest in the Himalayas really is the tallest mountain on the planet, measured above sea level.

I continued my journey northwards along the east coast to a part of the island that was much less populated than the south. This area was more heavily forested and here one could still find the Ifil, the official tree of the island, as well as Guam's official bird, the Totot, or fruit dove. Unfortunately, this was also the territory of a terribly invasive species known as the Brown Tree Snake, a slightly venomous species of reptile that is a native of the Indonesian archipelago, New Guinea, and northern Australia. The Brown Tree Snake was accidentally introduced into Guam in the twentieth century and it rapidly ate the native bird population into extirpation, or extinction, including the Totot, which can no longer be found on Guam.[1] Although the Brown Tree Snake was common in Guam when I visited the island in 1989, I did not see any snakes. But then again, I did not hike too deeply into the interior jungles either, and that is where they are most likely to be found. Guam is the most developed island in Micronesia and it serves as the transportation and communications hub for the entire region so there remains a real ecological concern about the possibility of inadvertently transporting the invasive Brown Tree Snake to other islands as a stowaway on an aeroplane, or ship. If that happens, other islands could also lose their unique bird life, as Guam has. Special measures are therefore taken in Hawaii and particularly on Oahu to prevent this species from becoming inadvertently established in the Hawaiian Islands.

Much of the land at the north end of Guam is devoted to military uses, principally Andersen Air Force Base, which is a major component of the United States force projection infrastructure in the North Pacific. At the time of my visit to Guam Andersen was reputed to be the repository of up to 450 nuclear bombs that could be carried in the bellies of the B-52 strategic bombers based at Andersen. Sometimes while I wiled away the hours on Fujita Beach I would see the B-52s in the sky, either taking off or returning to Anderson at the north end of the island.

After driving as far north as I could I turned back onto the west side of the island and drove south. I stopped to visit Two Lovers Point, or *Puntan Dos Amantes*, as it is called locally. The point consists of a very high cliff – one that was quite visible to me from Fujita Beach, and I was intrigued by the legend associated with it. Apparently, in the long past, two Chamorro lovers, a young man and a young woman, entwined their hair together and leapt to their deaths from the cliff after being hunted down by a Spanish sea captain to whom the girl had been promised by her parents. On the landward side of the cliff, there is also a deep basalt cave that drops all the way down to the ocean.

The view from Puntan Dos Amantes on Guam

From *Puntan Dos Amantes* I drove south back to Tumon Bay and the Fujita hotel. Tumon Bay was a major venue for both hotels and nightclubs in Guam. There was only a small beach bar at the Fujita hotel and it catered mostly to the young US mainland expatriate crowd although I found it to be a pleasant spot for a sunset or early evening drink. One evening I visited the Pacific Star hotel in Tumon Bay, which was not too far from the Fujita Hotel, and there I met a charming Chinese-American flight attendant for one of the major US airlines that served Guam who demonstrated an unusual skill. She was able to take one of those small plastic straw-like stir sticks that came with drinks and after placing it into her mouth she could tie it into a knot using only her tongue!

Hotels and resorts on Tumon Bay, 1989

My time in Guam was coming to a close but there was one more thing that I wanted to do. I had not done any snorkelling in Guam and I wanted to make at least some effort to see the underwater world of Guam. There was at that time a recent addition to the tourism options in Guam in the form of a small passenger submarine known as the *Atlantis* submarine. I was interested in this vessel not only because I thought it would be really neat to travel in a submarine, but also because I had learned that these recreational submarines were built by a Canadian firm out of Vancouver, British Columbia, and I was interested to learn more about a submersible that my own countrymen had created. With a large group of Japanese tourists, I was ferried out to the submarine, which was tied up at a buoy, and we boarded the vessel through a vertical hatchway and descended a set of stairs that led to parallel rows of seats next to large 60-centimetre viewing ports. There were two banks of seats on either side of the 20-metre-long vessel accommodating a maximum of 46 passengers, and each seat had its own large viewing port. There were three crew members, one a hostess. I took a seat on the starboard side and the other seats were rapidly filled by the Japanese tourists. The submarine was operated by electric propulsion and we soon slipped smoothly below the surface of the ocean, jugging along at one and one-half knots underwater as our hostess gave us a running commentary in English and in Japanese. Our submarine was rated for a depth of 45 metres and we probably achieved something close to that before emerging from the Pacific. During our short undersea voyage, we saw a variety of tropical fish native to Guam such as Moorish Idol, Unicorn Fish, Giant Trevally, Clownfish, Surgeon Fish, and even a sea turtle that swam by close to our submarine.

The Atlantis submarine

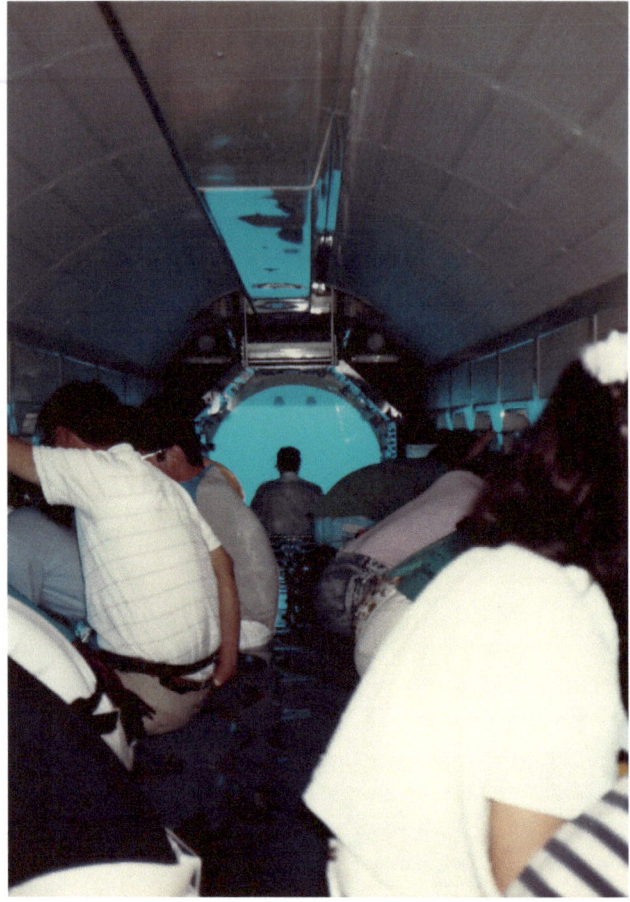

Tourists watch for marine life as the Atlantis cruises underwater

My journey to Micronesia had now come to end, as the sun set at Tumon Bay on my last night in Guam. On the morrow, I would re-board Continental Airlines Trans-Pacific Service heading east to San Francisco before making connections north back to Canada. I enjoyed a last drink at the beach bar at the Fujita. The next day I left Guam on Continental Flight Number Six, part of that airline's Trans-Pacific Service. Upon our aeroplane's arrival at Honolulu in the still-darkened early morning hours on February 24, 1989, I discovered that we had come to Honolulu in the aftermath of a travel tragedy. The airport was largely deserted as our aeroplane taxied towards the terminal but there was one aeroplane nestled against the terminal building that immediately drew the attention of the passengers sitting, like myself, on the right side of our aeroplane because it looked very peculiar in the dim light. There was a dark patch on the side of the aeroplane that slowly became visible as a yawning gap in its fuselage as our plane taxied nearer to it. When the passengers from my flight disembarked at the Honolulu terminal we learned of a tragedy that had occurred earlier that night.

The aeroplane parked on the tarmac as we arrived with a giant hole in its side was United Airlines Flight Number 811, which had been scheduled to fly from Honolulu to New Zealand earlier in the night but after its departure from Honolulu, the aeroplane had quickly encountered trouble as it rose to its cruising altitude. According to later press reports, the passengers on that flight remembered hearing a hissing sound and then a tearing, or ripping sound, as a large piece of the fuselage of the aeroplane was torn away, creating a huge three-by-twelve-metre vertical tear in the forward baggage compartment that also penetrated into the passenger cabin. Nine passengers who were still strapped to their seats were sucked out of the aeroplane to their deaths – some of them instantly, when they were sucked into the aeroplane engines that subsequently caught fire, while other passengers dropped through the air for four minutes until they hit the ocean and died from blunt force trauma. It was a horrific tragedy that resulted from a defect in the design of the doors to the baggage compartment, which did not lock securely. The decompression that resulted from the tear in the fuselage was so strong that one female passenger later reported that it had sucked her earrings right off of her ears. Naturally, the tragedy left me shocked by what had happened to the United passengers and it certainly gave me a great sense of unease, particularly because I had been flying on my own journey across the Pacific on large jumbo jets. Later, when my next flight arrived in Francisco were met by a bevvy of reporters asking what we had seen and learned of the disabled aeroplane in Honolulu, and a few of my fellow passengers spoke to the reporters from the print media and television crew about it.

Witnessing the results of what had happened to United No. 811 was a sobering reminder that life is all too short and that there are both risks, as well as rewards, in international travel. I thought of all those lives that had been cut short by the disaster on Flight 811, and a bit of rhyme was recalled to mind from my memory that spoke to lives cruelly cut short: 'Your expected future will pass from you like the sunlight from a bright field before the sweep of shadows, your past will become the keepsake of ghosts.'

NOTES

[1] As a result, the Guam legislature recently decided to drop the Totot as the island's official bird in favour of another species, an endangered rail.

PART III

RETURN TO THE LAND OF THE KAMEHAMEHA'S

13

OAHU AND HAWAII ISLAND

Oahu

Many years passed since my first visit to Hawaii in 1987, and I was not able to visit the islands again until the 2020s. In the interim, I married, raised a family, and worked through a professional career until my retirement. When I finally returned to Oahu in 2020, I found the island to be remarkably the same in some respects; in others quite different. Any seediness I had witnessed in Waikiki in the Eighties was now long gone, as a rather garrulous night clerk at the Hilton Hotel made a point of telling me when he learned that I had last been in Waikiki in the Eighties. Now I found that down-market bars and strip clubs had been replaced by chains of high-end boutiques catering to the tourist trade, including such high street retailers as Coach, Gucci, Chanel, and Hermes.

No longer was there any worry on the part of the locals in Honolulu that the Japanese would buy up the housing in the islands because the collapse of the Japanese economy in the Nineties had forestalled any major economic threat to the US economy from Japan. But that did not mean that all was well in the local real estate market. Although Japanese investment was no longer distorting the local real estate market Hawaii was now seeing an influx of money and people from the US mainland, and it was now wealthy continental Americans who were imperiling the dream of home ownership for the islanders. The overall cost of living in Hawaii, and especially the high cost of housing, has now forced many people born and raised in Hawaii to relocate to the continental United States. And an effort to implement a homesteading land distribution system for indigenous Hawaiians that was originally conceived by Prince Jonah Kuhio Kalaniana'ole, a member of the former royal family who served as Hawaii's congressional representative early in the twentieth century, has failed to satisfy the hunger for land amongst the remaining indigenous population. In fact, in 2020, indigenous Hawaiians were now among the poorest social class in what had once been their own country.

Although tourism remains of great economic importance to Hawaii, not least because of the employment it provides to multiple social classes, the exponential growth in tourism since I visited in 1987 has caused many islanders to reassess the social value of tourism. In the winter of 2020, on the cusp of the great Covid-19 pandemic, I found myself amidst a veritable tidal wave of tourists on Oahu. When my wife and I hiked to the top of Diamond Head we were part of a huge single-file, non-stop crowd on the trail to the rim of the crater. The streets of Waikiki were crowded with tourists, even quite late into the evening. For many on the islands, the real problem with tourism is the emphasis on mass-market tourism. Vast numbers of people are encouraged to come to the islands to visit the beach, imbibe tropical drinks, and attend an inauthentic tourist luau experience with a finale that is usually a spectacular fire dance – from Samoa! Mass market tourists remain oblivious to Hawaiian history, and to what remains of Hawaiian culture, while their impact on the natural environment creates pressures on the fragile Hawaiian ecosystems. Thus, growing voices among the local people have called for slower growth in tourism, if not an outright reversal of mass-market tourism. There are also calls for new measures to try to encourage tourists to undertake a deeper immersion into Hawaii than what is on offer in the standard tourist packages.

Single file trekking on the Diamond Head Monument Trail on Oahu in 2020

A Samoan-style Fire Dance demonstration ends many tourist luaus in Hawaii

For the tourist, the only negative experience is the widespread homelessness in Hawaii. Some of that is local, perhaps including some ex-military types who fell off the edge of society after their military discharge but in talking to taxi drivers I discovered that the greatest cause of the problem has been an influx of homeless people from the American mainland. That is a consequence of another downside to mass market tourism – cheap airfares that make it quite affordable for virtually anyone to wash up on Hawaii's shores where, if they are homeless, they can live without facing the fear of freezing to death in a continental winter. Poor mental health is an obvious contributing factor to the homelessness problem and a pointed reminder of America's increasingly frayed social safety net and healthcare system. The homeless problem in Oahu has been exacerbated by the lack of enough shelter space for the homeless on Oahu, although some of the homeless preferred to sleep rough and did not want to go to a shelter at night. I saw a large number of homeless men congregating along the Waikiki beaches at night, requiring police foot patrols to prevent any conflicts between the homeless and tourists. Some of them exhibited mental health issues, including one homeless man that I saw openly defecating next to a tree along crowded Kalakaua Avenue in the evening, horrifying a group of young American women walking just in front of me who admonished each other to 'walk faster' at the sight.

Mass-market tourism brings beachgoers to Waikiki on Oahu

The changes that Hawaii had undergone in the more than thirty years that had elapsed since my prior visit had a visceral effect on many of the long-time local residents, particularly those who were able to trace some indigenous blood in their genealogy. Many of those who possess mixed ancestry have joined with the few remaining pure indigenous Hawaiians to seek to recapture some of the customs and culture of their forebears. A Hawaiian renaissance dating from the late twentieth century has led to increased interest in Hawaiian history and culture among those who claim some indigenous ancestry, and that has also led to the creation of a Hawaiian sovereignty movement that draws upon the knowledge that a separate Hawaiian nationality once existed in these islands. Visible evidence of this new wave of thinking could be seen in the popular display of the Hawaiian flag, which dates from the period of Hawaii's status as an independent nation. I found that the Hawaiian flag was now much more prominently displayed in the twenty-first century than it had been during my visit in the twentieth century. Furthermore, I often saw it displayed upside down as an internationally recognized signal of distress, and as a new reflection of the uncompromising stance of the Hawaiian sovereigntists that has grown out of the cultural renaissance of *fin de siècle* Hawaii.

Hawaiian nationalists have gone so far as to attempt to regain some form of international status for Hawaii, notably in a 2001 international arbitration case called *Larsen v The Hawaiian Kingdom* that was brought before the Permanent Court of Arbitration at The Hague, in the Netherlands. Hawaiian sovereigntists sought to conjure up some form of continuing international recognition for an independent Hawaii based on the idea that Hawaii was an occupied country. To avoid having the US Government block the international arbitration the Hawaiian sovereigntists created a novel dodge by establishing a straw-man litigant against which the sovereigntists could file a case at The Hague. Thus, a so-called Regency Council was brought into existence to represent the defunct Kingdom of Hawaii. The sovereignists then claimed that those same regents had failed to take the necessary steps to end the US occupation of the islands. Sovereigntists were enthralled by the raising of the Hawaiian Flag, as a national flag, at the Hague court at the outset of the arbitration according to the established practice of the court but it was all smoke and mirrors because the Permanent Court of Arbitration correctly perceived that the litigant's real dispute was with the United States, and the United States had refused to become a party to the case. Absent the presence of the United States, the official parties had no real dispute to litigate and the case was dismissed. In legal terms, the Larsen arbitration may have been tantamount to tilting at windmills but the sovereigntist efforts at least sustained the hopes of those who wish to see the restoration of Hawaiian sovereignty someday. Most Americans consider that to be a fantasy. In the meantime, the US government has held itself prepared to offer indigenous and mixed-indigenous Hawaiians a status similar to that offered to Native American tribes residing in reservations on the mainland. So far, that is an offer that has not been taken up by the Hawaiians.

Princess Kaiulani, the last heir to the throne of the Kingdom of Hawaii, owned a beautiful estate in Waikiki called Ainahau long before Waikiki became the tourist mecca that it is today and, by coincidence, my hotel in Waikiki was located on lands that once formed part of her estate. On this trip, I delved into the life of this Princess, the memory of whom I had first encountered when I visited Hawaii in the 1980s. Her life was bound up with a last effort to sustain Hawaiian independence in the late nineteenth century. Immediately after the overthrow of the Hawaiian monarchy in 1893, Princess Kaiulani went to the United States to make public and private appeals to forestall the American annexation of her kingdom. Although she obtained some temporary success by delaying Hawaii's annexation the trajectory of American power made the acquisition of the strategically located Hawaiian Islands essential in the minds of Washington policymakers. Ultimately, despite her best efforts, the United States took the islands in 1898 during the Spanish-American War. I followed the course of Princess Kaiulani's life through the Hawaiian State Archives, and the Bishop Museum Archives, where I examined some of her original correspondence as part of my research for a book that I was planning to write on her efforts to save the kingdom.

The popular princess had a delicate constitution and she died within a year of Hawaii's annexation by the United States in 1898, but so great was the public affection for her, and the respect for the efforts she had made to save the country, that her political adversaries within the white-dominated government of the Republic of Hawaii felt compelled to honour her with a royal state funeral. She was interred at Mauna 'Ala, the Royal Mausoleum of Hawaii. A guide at Queen Emma's Summer Palace, which I visited on this trip with my wife encouraged me to visit Mauna 'Ala by telling me that very soon, when the last grandchild, and the last grandnephew, of the last two Hawaiian princes, had passed away the Kalakaua dynasty tomb will be closed and filled in with earth, as was the Kamehameha dynasty tomb early in the 20th century.

So I went to visit Mauna 'Ala and descended into the tomb of the Kalakaua dynasty. The royal remains of the principal members of the royal family are all there – the King and his Queen, and Queen Liliuokalani and her heiress, Princess Kaiulani (as well as both of Kaiulani's parents). It is a sombre place but the interior of the tomb is brightened by white marble facings. There is a white marble bust of King Kalakaua, the founder of his dynasty, at the far end where his tomb lies. Naturally, the tomb of the King who was the founder of the dynasty is the most impressive. The King's bust was heavily decorated with flower leis when I paid my respects. I found Kaiulani's resting place near the front of the tomb lying to the lower right with the simple inscription 'H.R.H Kaiulani 1875-1899.' Mauna 'Ala is a peaceful place behind an iron fence that bears the royal coat of arms. The Hawaiian flag flies alone here – one of the very few places in Hawaii where, by law, the Hawaiian Flag does not have to be paired with the American flag and so Mauna 'Ala lingers on in quiet memory as a last remnant of the lost Hawaiian Kingdom.

Princess Kaiulani in London in 1895 (Wikimedia Commons)

After Princess Kaiulani's death, the white minority government of the republic of Hawaii found it necessary, if not desirable, to memorialize her in some way and a decision was made to name a public school after her. Her namesake school still exists, although the neighbourhood that hosts the Princess Kaiulani School is now one of the poorest areas of the city of Honolulu, and the school is one of the lowest ranked in academic achievement in Honolulu. But her school has a unique artifact from her life – a direct descendant of her famous Banyan tree. During her lifetime Kaiulani's estate at Ainahau boasted the first, and for a time the most glorious Banyan tree in the Hawaiian Islands, and it was under her Banyan tree that the young Princess would sit to be regaled by tales told to her by a family friend, the famous writer Robert Louis Stevenson, author of *Treasure Island*, and other favourite adventure stories. When the Princess Kaiulani School was established the school officials approached her father, ex-Governor Cleghorn, and asked him for a transplant from Kaiulani's Banyan tree to grace the new school. Governor Cleghorn complied and a sprig from the original tree was provided to the Princess Kaiulani School and planted on its grounds. That tree grew into a tall Banyan tree in the far corner of the school field, where it remains today.

As for the original tree at Ainahau, the victorious annexationists saved it for a time after the estate was sold, although they always referred to Kaiulani's Banyan tree as Stevenson's Banyan tree, despite the fact that the tree had been located on Princess Kaiulani's property. More recent historiography now correctly identifies the original tree as Kaiulani's Banyan Tree. Eventually, the original tree was cut down as a hazard and today only the Banyan tree at the Princess Kaiulani School remains as a living link to the original Banyan tree from Kaiulani's Ainahau estate at Waikiki. It flourished at the school until recent years. Regrettably, a few years before my latest visit the tree was severely wounded by a fire that was deliberately set by an arsonist. But the tree lived, and it is now surrounded by a fence, as of 2020. Intriguingly, during the efforts to put out the fire set by the arsonist, some observers claimed to have seen an apparition of Princess Kaiulani within the smoke and haze of the fire.

The schoolyard clone of Kaiulani's Banyan tree

On this trip, I was also able to visit Washington Place, which had been the private residence of Queen Liliuokalani prior to her death, and afterward, it became the official residence of the territorial and state governors. It is now a museum, albeit one that is still heavily guarded, and requires a pre-authorized entry because of its close proximity to the new Governor's Mansion built nearby. Washington Place remains both architecturally and historically important in Hawaii. The first floor has been carefully restored to its appearance when Queen Liliuokalani lived there following the collapse of the Hawaiian monarchy and many of her original furnishings still remain. The upper floor, however, has been totally altered over the years by successive governors and their families and now has no substantial historical value. It is the venue for a small museum about the house and the monarchy, and it contains several interesting artifacts, including a well-known flower painting that was done by Princess Kaiulani as a gift to the Queen.

Washington Place in Honolulu, the historic home of Hawaiian royalty and American governors

Other important sites in downtown Honolulu include the Mission House Museum, which tells the story of the early Congregationalist missionaries who came to Hawaii from New England beginning in the 1820s. The site preserves the early structures built by the missionaries and the various devices used by mid-nineteenth-century settlers. An early printing press reminds visitors of the important efforts made by the missionaries to create a written Hawaiian language for the indigenous inhabitants. The Congregationalist missionaries were both a shield and a bane for the emergent Hawaiian kingdom in the early 19th century. The missionaries helped the early kings to establish a form of government that was able to obtain the respect and recognition of the major western powers and the missionaries greatly accelerated the education and adoption of modern thinking among the Hawaiian population. However many of the descendants of the early missionaries subsequently built large commercial empires in Hawaii, and they essentially took command of the Hawaiian economy. Later still, some of them conspired against the monarchy and they formed the backbone of the annexationist movement that eventually wrested political control of the country away from the indigenous population in the 1890s, before turning the country over to the United States.

The Mission House Museum in Honolulu commemorates the Congregationalist missionaries

The heart of the Congregationalist power and influence in early Hawaii was centred on the Kawaiaha'o Church, which is also known as the royal church of Hawaii. Members of the royal family initially favoured this church and it became the setting for important religious events in the lives of the members of the royal families, even after some of them joined the Anglican Church. It is an old and historic building, and it also is the site of a mausoleum dedicated to King Lunalilo, who reigned briefly between the Kamehameha and Kalakaua dynasties, and who refused to be interred in the royal tombs at Mauna 'Ala. The Kawaiaha'o Church fell out of favour with the royal family during the reigns of King Kalakaua and Queen Liliuokalani. Its pastor publicly opposed Queen Liliuokalani and sided with the white minority in the aftermath of the overthrow of the monarchy, which resulted in an exodus of indigenous Hawaiians from the Congregationalist Church. Today, however, the church still conducts some of its contemporary services in the Hawaiian language.

The remainder of Oahu speaks to the dominant American cultural influence of the present. My wife and I made the essential visit to Pearl Harbour, where I found the exhibits improved from my earlier visit in 1987. A new feature, now juxtaposed to the Arizona Memorial Pavilion, is the *USS Missouri*, the 'Mighty Mo,' which was one of the US Navy's most powerful battleships in World War Two. *Missouri* forms a nice counterbalance to the wreck of the *USS Arizona*, and the two ships, one a wreck, draw a direct link between America's defeat at the Battle of Pearl Harbour and America's ultimate victory in the Pacific War. It was on the deck of *Missouri* that the Japanese formally surrendered to the United States and its allies in Tokyo Bay on September 2, 1945.

The USS Missouri is now a floating museum at the Pearl Harbour Naval Base

We toured the great battleship and we were able to stand at the actual site of the Japanese surrender ceremony. There was an interesting Canadian story about the surrender ceremony. The original surrender document was signed imperfectly by the representatives of the allies because the Canadian representative, Colonel Lawrence Cosgrave, was blind in one eye as a result of an injury he suffered in World War One. That caused Cosgrove to sign his name on the wrong line on the Instrument of Surrender, thus forcing each succeeding allied signatory to sign their names in the wrong place. General MacArthur's adjutant tried to fix the problem by striking out the typed names and writing in the correct names under their actual signatures on the document but the punctilious Japanese considered the original document to be unacceptable, and it had to be re-signed privately after the ceremony was over. *Missouri* displayed copies of the original defective version and the subsequent corrected version of the surrender document that ended the most devastating war of the twentieth century.

Afterward, we toured the interesting Pearl Harbour Battle Museum, which contains artifacts and models of the ships that participated in the battle. Naval history has always fascinated me and I took my time reviewing the exhibits on display. But after a while, my wife became restless because we had come to Pearl Harbour on an organized bus tour and she noticed that fewer and fewer people that she recognized from our tour were still in the museum. Eventually, she left for the bus without me after extracting my promise to quickly finish up and rejoin the tour. When I finally reached the bus I was the last person to rejoin the tour and the whole bus erupted in applause as I came aboard! The moment made me realize something about visiting the site of the

Battle of Pearl Harbour that now differed from my first visit. When I first came here in 1987, the battle was not only history – it was also a memory. Many of the people who walked the site with me then remembered the battle but in 2020 those people were gone and for the tourists of the early twenty-first century this was only history – important to be sure but something remote and no longer personal as it was to an earlier generation of visitors.

On another day we drove out to the Dole Plantation – a lasting tribute to the pineapple, which once was a mainstay of Hawaii's agricultural economy but that has now disappeared, owing to the cost of doing business in Hawaii. The founder of the Dole Pineapple Company was James Dole, a relative of Sanford Dole, the first and only President of the Republic of Hawaii which replaced the Hawaiian monarchy in the 1890s. Educated at Harvard, and trained in business and agriculture, James Dole realized that his connections in Hawaii could profit him and he arrived in 1899 with a view to setting up a pineapple operation. His venture was a success although, eventually, his company passed into other hands. Today the company remains well-established in the United States although it no longer grows pineapples commercially in Hawaii. After statehood was achieved in 1959 the cost of labour in Hawaii rose and it eventually became more economical to shift pineapple production to Central and South America. The Dole Plantation is located in the original Dole company fields where, in the Fifties, it originally served as a fruit stand. In 1989, the site was rebuilt as a plantation-style structure to serve as a museum and tourist experience revolving around the history of the pineapple in Hawaii. Some pineapples are still planted and harvested on-site for the tourist operation. Though it was very much oriented toward the mass-tourism market we found that it was a nice place to visit, especially for families. There was a short train ride around the grounds that gave a good overview of the gardens and fields surrounding the main pavilion, including some of the fields where pineapples are still grown as a tribute to the one-time importance of the fruit to Hawaii.

The Dole Plantation on Oahu

Oahu still remained a very special place in my mind during my second visit but I could not ignore the fact that there are some obvious fissures in Hawaiian society, particularly emanating from those who claim some indigenous ancestry, and who contest American domination of the islands. Americans from the mainland view the historical relationship between the United States and the islands quite differently. Both elements, however, seem to agree that mass-market tourism, particularly on Oahu, will probably impact the environment and the quality of life that all local residents sought, and expected, in a negative way. Climate change, environmental degradation, land costs, inward migration, the importance of tourism, the sovereignty movement, and the role and responsibilities of the many military institutions within this American fortress island all mix into a socio-political brew in the early twenty-first century. It remains to be seen how these emergent tensions and contradictions will be reconciled on this island.

The Beach at Waikiki on Oahu

Hawaii Island

I had long wished to visit the island of Hawaii, the namesake of both the former country and the US state of the same name in part because it was the island that had seen the birth of the most consequential leader of the Hawaiians, Kamehameha the Great, the founder of the Kingdom of Hawaii, and its political unifier. The Big Island, so-called by the residents of the state of Hawaii, is also the most ecologically complex of the islands in the Hawaiian chain. It is by far the largest of the Hawaiian Islands, having a total area of 28,311 km. It is also the youngest and most volcanically active of these islands. It boasts ten of the world's fourteen recognized climate zones, including a polar tundra zone on the tops of the mountains of Mauna Kea and Mauna Loa. It was on the southernmost tip of this island that the original waves of Polynesian settlers, voyaging from the Marquesas and Society Islands far to the south, first made landfall on the then uninhabited Hawaiian Islands. I was eager to visit the island, and encouraged by the onset of winter as 2021 passed into the new year of 2022 I made plans to visit the island with my wife Vi.

But then there was Covid-19. The pandemic disease that erupted in China in 2020, still disrupted foreign travel two years later, and it presented real health risks to travellers and sedentary folk alike. My wife and I debated whether we should risk a foreign trip with the pandemic still raging. Not only were there the dangers posed by the disease but as well there was the need to contend with the substantial travel barriers that various governments had erected both for outward-bound travel as well as return travel. In the end, we decided to go because we felt that since the world had failed to eradicate the disease, Covid should now be treated as an endemic disease like many others that we humans encounter. We began to think of adapting ourselves to living with Covid, rather than shutting ourselves away from the world in hopes of avoiding it. By late 2021, vaccines against the disease were widely available and before travelling outside of Canada we took steps to ensure that we were fully vaccinated with two full shots of Moderna's Covid-19 vaccine and an additional half-shot booster before we travelled. American travel restrictions also required that we obtain a negative Covid-19 antigen test within forty-eight hours prior to our flight.

Travelling abroad during the time of Covid was a slightly surreal experience. We left our home in Ottawa on a cold, wintry, and snow-covered Sunday morning in January beneath a darkened sky. The Air Canada check-in counter at the airport was slow but the older gentleman behind the counter helping us managed to find two seats side-by-side on our initial flight to Vancouver, British Columbia. Once aboard our Air Canada flight, we found that our aeroplane was jammed with travelers, and every seat was filled despite the pandemic. Clearly, we were not the only people unafraid to travel at this time. Aboard the aeroplane everyone wore medical masks, a requirement imposed by the government and by the airline because of the pandemic. I chose to wear an N-95 industrial mask because it offered maximum protection but the tight fit of my industrial-grade mask grew increasingly unbearable as the day wore on, and I was glad when we arrived in Vancouver because I was able to remove the mask and escape from the airport for a brief period to breathe the fresh air outside.

We had a very long wait for our connecting flight to Kailua-Kona on the island of Hawaii and it was a full seven and one-half hours before we boarded the Air Canada flight at Vancouver that would take us to Hawaii Island. However, the flight to Hawaii Island was much more pleasant than our domestic flight from Ottawa if only because the Hawaii-bound aircraft was filled to about one-third of its total capacity, thus giving us plenty of room to stretch out (and avoid other travellers).

I was struck by the great pains taken by the Air Canada cabin crew on this flight to protect themselves and their passengers from Covid-19. When delivering food or drinks they wore masks and rubber gloves and the sole female flight attendant present on the flight appeared to be gowned like a hospital worker. The cabin crew kept periodically reminding everyone over the aeroplane's public address system to keep their masks on at all times unless passengers were eating or drinking, and even then passengers were cajoled into finishing their repasts within a period of fifteen minutes.

Travelling during the time of Covid was more than a bit unnerving, and one could not help having second thoughts about the wisdom of the travel choices we had made even though we knew now that learning to adapt to the disease was necessary. On this flight, I found that my N-95 industrial mask was simply too uncomfortable to wear because it became too difficult to breathe when I tried to sleep during this night flight. It got to the point that I could not sleep at all on the plane because every time I nodded off I would be awakened by a growing panic for air. Finally, I tore the N 95 mask off and resorted to one of the less effective but more comfortable blue medical-surgical masks. In any event, the risks of contracting Covid were much less on this particular flight because it was below capacity and because we were flying on a new B737 Max aircraft that had the most modern and up-to-date ventilation system.

Finally, we landed at Kona. After obtaining our luggage we hired a taxi and went off to the Royal Kona Resort for two nights. The desk clerk who met us was a very pleasant woman who gave us an upgrade to a corner room overlooking the ocean. Afterwards, we just sat outside and watched the glittering reflection of the moon on the ocean waves as the breaker waves rolled in and beat against the rocky shoreline. In the warm Hawaiian night, we experienced a magical tropical vignette that banished any arduous memories of our aeroplane journey.

Sunrise at Kailua-Kona on Hawaii Island

The next day we rose early to the sound of the breakers roaring against the rocks below, while beyond the breakers the broad Pacific Ocean beckoned to us in all its blue glory. We enjoyed breakfast in the hotel restaurant overlooking the ocean. Afterwards, we explored the grounds of the resort and then we walked into Kailua itself. The town site of Kailua seemed a little tacky and touristy, a sort of down-market Waikiki. The west side of the island is clearly the centre of tourism, however, and Kailua-Kona is the epicentre of west Hawaii Island tourism. Hilo, the other major centre that is located on the east side of the island is more diverse and offers a less frenetic form of tourism. Despite our initial impressions of Kailua however, I noted that there was a picturesque, and partially shaded, walk along the shoreline behind a seawall that passed by some significant points of historical interest including Hulihee Palace, a sanctuary for royalty, and the Moku'aikaua Church, which was founded in 1820 by New England missionaries, and which is now the oldest extant church in the Hawaiian islands.

The seashore walk in Kailua, Hawaii

The stone Moku'aikaua Church, built in 1836 in Kailua

Later, we relaxed back at our hotel and in the evening we ate a meal on our balcony overlooking the Royal Kona's Hawaiian Luau Show below us. The hotel's luau provided attendees with the typical spectacle, including lots of Hawaiian music and dancing, some of it historically authentic, although much of the tourist-grade hula is a watered-down version of the original. It ended with the spectacle of the Samoan fire dance, as do most tourist luaus in the islands, even though the fire dance was never part of the ancient Hawaiian culture, and actually comes from the islands of Samoa in the South Pacific.

An authentic luau is both a cultural and a culinary event. The traditional food served at a Hawaiian luau is Kalua Pork, a dish prepared by stuffing a pig's carcass with hot stones, wrapping it in ti leaves, and cooking it for twelve hours in an earthen oven called an *imu*. This process gives the pork a smoky taste. Pork, or fish, is also served wrapped in taro, which is called Lu'au leaves and that dish has become the namesake of the entire feast. Another favourite dish is Lomi Lomi salmon, a salted salmon dish that originated when salted salmon was imported into the islands in the mid-nineteenth century by the Hudson's Bay Company from its posts on the North American continent. And always at a luau, poi is served. Poi is a bland dish made from the taro plant but it has been the staple of Hawaiian cuisine since time immemorial.

Another Hawaiian staple, and one that I much appreciated, is Kona Coffee. Kona coffee is a world-renowned coffee bean and is considered to be among the finest coffees. With a lighter and fruitier taste that is not too acidic Kona Coffee has obtained devoted adherents all over the world. Sometimes it is blended with other coffees from South or Central America, and the true Kona taste is lost but efforts have been made in Hawaii in recent years to protect the purity of the brand. The Kona coffee farms are found on the west side of the island in the uplands away from the coastline above Kailua-Kona and they tend to be small, family-owned farms. There are about 700 coffee farms on Hawaii Island producing the island's unique blend of coffee.

I rose early the next day and once again took in the spectacular views of the Pacific Ocean from our hotel balcony. Afterwards, I headed to Kailua to see two historical sites of interest to me. At the King Kamehameha Hotel, I examined the reconstruction of the Ahu'ena Heiau, the sacred venue where King Kamehameha I, also called the Great, and sometimes also called the Napoleon of the Pacific, spent much of his time as ruler after he completed the political unification of the islands. Here, in the last years of his reign, he held court and attempted to navigate the difficult path of reconciling the indigenous customs and traditions of Hawaii with the new customs and traditions resulting from western contact. While Kamehameha the Great held fast to the old Hawaiian ways during his reign he also readily adopted the fruits of western technology, both in marine transportation and in military affairs, which gave him the advantages he needed in his wars of conquest to unite the islands into a single kingdom. Although all of the current structures of the Ahu'ena Heiau are reproductions of the original buildings and structures, the site of Ahu'ena Heiau is still held to be sacred by the indigenous Hawaiians, particularly since it was at this place that King Kamehameha the Great passed away peacefully. From here his body was prepared for its secret burial, according to traditional Hawaiian customs.

No one knows for certain where the remains of King Kamehameha the Great lie on the island of Hawaii. His courtiers secreted his remains in an unknown place so that no one could seek to draw strength from the late King's *mana*, or spirit, from his remains. But while I was visiting Hawaii Island I happened across an interesting article in the January, 2022, edition of the local magazine *Ke Ola*. In its pages was an article telling the tale of a local man, Tyrone Young, who had been hired as a guide for an American photojournalist retained by the National Geographic Society for a magazine article in 1983.[1]

The magazine article related the seizure of an American sailing vessel, the *Fair American*, and the massacre of most of its small crew in retaliation for an earlier public flogging of an important indigenous Hawaiian chief by the father of the young captain of the *Fair American*. The ownership of the seized *Fair American* was passed to the rising Kamehameha, who made the *Fair American* the centrepiece of his new Hawaiian war fleet. Included in the gift to Kamehameha were the cannons of the *Fair American and* its surviving crewman, Isaac Davis. With his new war fleet, and assisted by Davis and by John Young, a crewman stranded from another ship, Kamehameha conquered the islands of Maui and Oahu.

Kamehameha realized that the massacre of the *Fair American's* crew might have significant repercussions so he ordered the dead crew to be interred according to Hawaiian custom in a burial chamber in a cave-burial site. It was to visit that cave-burial site that Tyrone Young was engaged to assist a National Geographic photojournalist in 1983.

Upon entering the cave Tyrone Young found that there were three burial chambers in the cave. In the first chamber, he found a large canoe that famed Hawaiian artist and historian Herb Kane had asked him to measure. Young found that the canoe was twelve feet long but then, inside the canoe, was the full skeleton of an indigenous Hawaiian who Young measured to be almost seven feet in height – the reputed height of King Kamehameha the Great. Young, viewing the remarkable sight of the skeleton in the canoe, thought to himself that this must be the King, as what better place to hide his remains than in a crypt set aside for foreigners? Who among the indigenous population would have thought to have looked there for the King's *mana*? Whether accurate or not I thought it made for a great story and the *Ke Ola* magazine article continued the ongoing mystery surrounding the great King's final resting place.

After exploring the Ahu'ena Heiau in Kailua, I viewed a remarkable gallery of paintings by the Hawaiian artist Herb Kane that was contained within an annex of the King Kamehameha Hotel. Herb Kawainui Kane was a Hawaiian artist and historian whose works of art were based on careful historical research. His art helped to energize the Hawaiian cultural renaissance that emerged in the waning years of the twentieth century. Kane captured the mythology of the islands in expressionistic portrayal as well as in highly realistic art depicting the major historical events in Hawaiian history. I found his gallery at the hotel quite mesmerizing, and a tribute to the life of a vanished civilization in all of its splendour and depth.

In Kailua, my wife and I visited Hulihee Palace, the summer residence of the monarchs of Hawaii. At almost 200 years old it is by far the oldest royal residence in the Hawaiian Islands and the only extant palace that was used by both the members of the original Kamehameha Dynasty as well as the later Kalakaua Dynasty. It is perhaps more accurate to speak of Hulihee as a mansion, or a villa, rather than a true palace. The Daughters of Hawaii, who now have charge of the estate, have managed to obtain the return of many furnishings that were once owned by the two Hawaiian Royal Families. Among the many interesting artifacts one, in particular, caught my attention. It was a music box once owned by Princess Kaiulani, a gift made to her by the writer Robert Louis Stevenson, the famous author of *Treasure Island* and other novels, who spent considerable time in Hawaii during the fading days of the monarchy.

One of the curious facts about Hawaii is that the remaining historical sites associated with the monarchy and the independent Kingdom of the nineteenth century are maintained for the public by private entities rather than by the state or federal government. Thus, private entities administer the seat of the royal government, Iolani Palace, in Honolulu, and the summer palace of Queen Emma, the last queen of the Kamehameha dynasty, on Oahu, as well as Hulihee Place, and the birthplace of King Kamehameha III on Hawaii Island. The US National Park Service can't administer these sites because it is impossible to fit the history of the Kingdom into the traditional American narrative without offending the indigenous Hawaiians. It is only the Polynesian archaeological and cultural sites of pre-western contact Hawaii that are shorn of politics and can thus be preserved, and administered, by the US National Park Service.

Hulihee, the residence of Hawaiian monarchs on Hawaii Island

After touring Hulihee Palace we repaired to the neighbouring King Kamehameha Hotel where we enjoyed a delicious treat of Hawaiian shaved ice. Afterwards, we left Kailua-Kona for Hilo, travelling around the north end of the island to reach Hilo, and experiencing both the arid west coast, as well as the lush east coast of the island before reaching Hilo just in time to view the departure from Hilo Harbour of a major cruise ship, the *Grand Princess*, one of the first cruise ships to resume Hawaiian service as the commercial effects of the Covid-19 pandemic slowly began to dissipate.

We remained ensconced in Hilo for the remainder of our visit to the Big Island. Our hotel room overlooked Hilo harbour and gave us not only views of cruise ships passing to and fro but also a view of a whale in the harbour and dolphins, including a spinner dolphin that jumped straight up out of the sea as I watched. We drove out to visit a pleasant waterfall within the city limits of Hilo itself, the 24-metre Rainbow Falls, or Waianuenue, that, on sunny days, produces rainbow colours in the mists as the water returns to the river after spilling over the falls.

The view from our hotel balcony at the Grand Naniloa in Hilo, Hawaii, 2022

One of the pleasant features of Hilo that was close to our hotel was the Queen Liliuokalani Gardens, a Japanese-style garden located on land along Hilo Bay that had originally been donated to the public by the deposed Hawaiian queen. The gardens presented us with a pleasant oasis, one that has been featured on one of a series of Hawaiian postage stamps issued by the US Post Office. We walked among Japanese-style stone bridges across ponds with small fish and crabs beneath Banyan trees.

The Queen Liliuokalani Gardens at Hilo, Hawaii

Adjacent to the park there is also a small garden dedicated to the American Bicentennial of 1976. It seemed to me that a garden dedicated to American independence in the eighteenth century appeared incongruous lying next to the gardens dedicated to the Hawaiian head of state deposed by Americans in the nineteenth century but then the indigenous Hawaiians and the Americans have had a complicated history.

We took a walk through the historic old town of Hilo and felt like we were walking through an old South Seas seaport of a century ago. Clapboard buildings reminiscent of the early twentieth century slowly gave way to Art Nouveau and Art Deco architecture. The Palace Theatre was most impressive and we stopped there for a drink and a snack. But perhaps the most striking structure was the old Federal courthouse which contained elements of Greco-Roman architecture mixed with an Oriental motif in the form of a tiled roof that made me think that the building might not have looked out of place in a Chinese seaport of the late 19th or early 20th century.

Historic downtown Hilo on Hawaii Island

We strolled through the public park on Coconut Island, which had been severely damaged in the 20th century by tsunamis. One tree on the island holds markers showing how high the tsunami waves reached during the episodic tidal waves of the past century. The marker for the 1946 tsunami wave reached 26 feet (7.9 metres) above ground level on Coconut Island, a fearsome spectacle in our imaginations.

White bands on a tree at Coconut Island show the height of tsunamis

Afterwards, I went for a long walk around the Grand Naniloa golf course perimeter and walked back to the old town to get a sense of how much damage the 1946 and 1960 tsunamis did here. The Grand Naniloa golf course itself is an example of the terrible effects of a tsunami. Once upon a time what is now the golf course was a thriving village within the Hilo town site but that village was destroyed by the 1960 tsunami and never rebuilt. There is a famous preserved clock on Kamehameha Avenue that is stopped at the exact time (1:05 PM) when the 1960 tsunami hit Hilo. Today, the clock is maintained as a public memorial adjacent to the perimeter of the Grand Naniloa golf course. Though rare, tsunamis remain a real threat in Hilo and the expansive green space along Hilo Bay (and there is a lot of green space) marks the areas within Hilo that were destroyed by tsunami surges in 1946 and in 1960. There is inequality in Hilo, as in much of America, and the homeless were present in the public parks along the waterfront, leaving them vulnerable if a tsunami were to strike.

And then, on Thursday, January 15, 2022, we awoke to the news that a tsunami advisory had been issued for all of the Hawaiian Islands as a result of a tremendous volcanic explosion in the Tonga Islands in the South Pacific. I immediately began watching for any unusual wave action from our balcony overlooking Hilo Bay and although I thought that the waters did seem a little more agitated than the day before there was no substantial wave action in East Hawaii. Later that day the tsunami advisory was cancelled. In Hawaii, there were only minor impacts resulting from a rise in wave action of about one metre above the normal sea level. But according to scientists who monitored this event, the explosion of the Hunga Tonga-Hunga Ha'apai Volcano in the Tongan islands was the greatest volcanic explosion on Earth since the Krakatau Volcano explosion in the Netherlands East Indies (now Indonesia) in 1883. The Hunga Volcano exploded with the force of a nuclear bomb, and it created a 5.8-magnitude earthquake. The sonic boom caused by the explosion was heard as far away as Alaska and in Canada's Yukon Territory.

Although the physical impacts in Hawaii were minimal the event was a sobering reminder of how vulnerable the Hawaiian Islands are to potential tsunamis, as we saw only too clearly during our recent visit to nearby Coconut Island. In an article that I read in the November, 2021, edition of *Ke Ola* Magazine a survivor of the 1960 tsunami in Hilo recounted his narrow escape from that tsunami, saying that he had run faster than he had ever run before, and even then he only just barely escaped the onrushing wave.[2]

Later in the day when we visited west Hawaii, we saw the only evidence of ruin caused by the minor tsunami waves that had reached Hawaii from Tonga. At Keauhou Bay, where we stopped to visit the birthplace of King Kamehameha III there was some damage where wave action had broken into a store near the shoreline. Some picnic tables were also broken up at Kahaluu Beach by the rising waves. There was some local criticism about the fact that a tsunami warning was only issued at 2 AM, and the tsunami wave had struck almost immediately afterward. Concern was also expressed that there was no accompanying tsunami siren activation. Despite those criticisms, the danger of tsunamis is taken quite seriously on Hawaii Island, and we often saw signs directing people to the location of evacuation zones in the event of a tsunami. The US government monitors the Pacific Ocean for earthquakes and volcanos that can cause tsunamis in the Hawaiian Islands, and scientists have even studied sand particles to determine the extent to which ancient tsunamis impacted the Hawaiian Islands to refine their assessments of the tsunami risks to the archipelago.

In the South Pacific, the volcanic eruption in Tonga was massive, with a towering plume of smoke and ash rising above the tropical archipelago and preventing any aeroplanes from landing or taking off. Communications with the remote island kingdom of Tonga were sundered for several days because the volcano destroyed the oceanic cables that connected Tonga to the rest of the world. Australia and New Zealand sent military overflights to Tonga to attempt to assess the impacts of the explosion. There was much devastation in Tonga, with homes destroyed and volcanic ash covering everything, and four people in Tonga were killed, as well as two in Peru, which also received a tsunami wave. Later, scientists determined that the volcanic explosion had sent giant atmospheric shock waves called Lamb Waves circling the globe four times in one direction and three times in the opposite direction. A pressure wave also shot up from the surface of the Earth all the way up to the Ionosphere, reaching the very edge of outer space some 450 kilometres above the Earth's surface. The changes to atmospheric pressure even caused increased wave sizes in the far-off Caribbean Sea.

To reach the west coast we drove across the middle of Hawaii Island, traversing the cross-island Saddle Road through hardscrabble volcanic terrain between the giant mountains of Mauna Loa and Mauna Kea. It was astonishing to see snow on both of those mountaintops in a Pacific island tropical paradise. The astronomical observatories on Mauna Kea are a subject of great controversy among native Hawaiians and they were clearly visible to the naked eye. Astronomy is both a bane and a boon to Hawaii. Because of its geographical position, and altitude, Mauna Kea, the highest mountain on Earth as measured from its base on the sea floor, is an ideal location from which to spy on the heavens. From the heights of Mauna Kea, astronomers can see 100% of the northern sky and about 80% of the southern sky with remarkable clarity. Even the Southern Cross constellation is sometimes visible despite Hawaii's position in the northern hemisphere. Consequently, through the management of the summit by the University of Hawaii several countries have invested in the construction of astronomical observatories on the summit. According to a published report during our visit[3] the astronomy undertakings on the Big Island resulted in spending of about 110 million US dollars throughout the state in 2019, and with indirect economic activity that increased to about 220 million dollars of spending, half of which was spent on Hawaii Island. The observatories and associated activities also supported 611 jobs on the island.

However, many people connected to the indigenous culture consider the summit of Mauna Kea to be a sacred site and the presence of the observatories on the summit to be a desecration of an important cultural site. During our visit, it was announced that a bill would soon be introduced into the state legislature to strip the University of Hawaii from its leasehold of the summit, and from all of its summit management responsibilities in favour of a new state agency that would have the power to restrict astronomical developments. A couple of days later the University of Hawaii's Board of Regents countered that development by suddenly voting in favour of a new Mauna Kea Master Plan, over strident local opposition, as the University sought to buttress its position as an astronomical site manager. At the Board of Regents meeting, submitters were reported as saying 'Aloha aina [love of the land] will defeat [the] University of Hawaii, aloha aina will defeat Manifest Destiny!'[4] The controversy has led to protests and even an interruption in access to the mountain in the past as the Hawaiian cultural renaissance and sovereignty movement increasingly collides with the objectives of modern society in the Hawaiian Islands. A particular lightning rod has been the proposed Thirty Metre Telescope, which has sparked both protests and litigation.

Along Saddle Road, amidst a volcanic 'moonscape' at a high altitude, one can also find unique islands of biological diversity. At certain points along Saddle Road, there are isolated environments called *kipuka* where lava flows from Mauna Kea and Mauna Loa worked around isolated segments of the Hawaiian forest. When the lava cooled, and hardened, islands of ecological diversity resulted that have remained isolated and protected from the effects of invasive species. Ten months after our visit to Hawaii Island Mauna Loa began to erupt again for the first time in forty years, giving rise to new lava flows that threatened the Saddle Road.

Unique species of Hawaiian insects continue to live in these *kipukas*, and they also remain a particular refuge for native bird species, including the wonderful I'iwi and the Apapane, both of which are bright red, or red and black and once gave their feathers to the makers of the renowned feather capes worn by Hawaiian royalty in the days of the Hawaiian kingdom. Now increasingly rare, these birds can still be found in protected zones at higher altitudes where the invasive avian malaria that has decimated the native Hawaiian bird life at lower altitudes cannot yet reach them. But with climate change affecting all environments, one wonders how long it will be before the remaining native Hawaiian bird species will be wiped out? Already Hawaii has suffered more species loss than almost anywhere else on the planet. Invasive species are a continuing threat to native Hawaiian flora and fauna. At one *kipuka*, the Kaulana Manu Nature Trail, there are at least seventy indigenous Hawaiian plant species extant and hikers are warned to brush their shoes before entering the *kipuka* in order to prevent an invasive fungal disease from decimating the local plants.

Avian malaria is a particular threat that was unknown in the islands until 1936 when it was first noted by Hawaiian pigeon fanciers, and the disease may have been brought to the islands by a conscious decision to import the non-native Pekin Robin, or Pekin Nightingale, through a public subscription that included school funding drives, all in a misguided attempt to add low altitude songbirds to the Hawaiian environment. Efforts are sometimes made to eradicate invasive species and, for example, programs have been devised on Oahu to prevent iguana lizards from becoming established. So far fruitless efforts have been made to disestablish the noisy coqui frog that was inadvertently introduced to Hawaii from Puerto Rico, and which is now well established on the eastern side of Hawaii Island.

We found a somewhat similar sanctuary to a *kipuka* during our visit to the sacred mountain of Mauna Kea. The mountain access road off of the Saddle Road is marked by Hawaiian flags and banners among the tents and detritus of the indigenous protestors who have, in recent years, periodically blocked the access road to the mountain to inhibit the further development of the astronomical complex on the summit of the mountain. However, at the time of our visit, the protestor's encampment appeared to be deserted. The road up to the visitor centre is quite steep, although it is paved, and it was possible for the two-wheel drive Chevy Camaro that I drove to reach the visitor centre without any difficulty.

The Mauna Kea Visitor Centre is located at an elevation of 2800 metres (9200 feet) and of course the air is much thinner at that altitude. Just reaching the visitor centre allowed us to take some photos that showed that we had ascended slightly above the clouds. We explored the visitor centre and watched a video explaining the purpose of the several astronomical observatories at the summit (including one sponsored by Canada). We could not mount to the summit itself, which is almost 4000 metres (13,000 feet) above sea level because only four-wheel-drive vehicles are permitted on the rough road that exists past the Visitor Centre. It was at the Visitors Centre that we discovered a small sanctuary for the rare indigenous Silversword plant and, as at a Saddle Road, *kipuka* visitors here are warned to brush their feet on the foot brushes that are provided before passing a gate into the small Silversword sanctuary. After our visit on the steep descent back to Saddle Road I had to keep the car in low gear most of the way to avoid burning up the brakes.

The following day we went off for a day tour of the beautiful Hamakua coast. *Hamakua* means 'breath of life' and we certainly found it in this Eden-like paradise. We stopped first at the famed Akaka Falls State Park to view the magnificent 135-metre waterfall that is set amidst a verdant glen. It is one of the most beautiful waterfalls in the Hawaiian Islands, and the most magnificent waterfall on this island. A legend says that this waterfall was named after a youth who threw himself over the falls to escape the hounding of bullies. As we explored the state park that contains the Akaka Falls I was surprised to learn that almost all of the plants here are non-native species that have been introduced to the ecosystem, and that have largely replaced the indigenous botanical specimens. The local fish are native to Hawaii however, and one of them is quite unique. A species of goby, known as the o'opu alama'o begins its life above the falls but drifts down the falls and through the lower streams to the ocean. When it is time to reproduce the fish returns to the pool below the Akaka Falls and then begins an arduous climb up the rock face behind the waterfall, using specially adapted fused pectoral fins on its underside that act like a sucker, and allow it to climb to the top where it spawns in its preferred pools.

The Akaka Falls on the Hamakua coast

The Hamakua Coast annually receives about 200 cm of rain each year making it the wettest part of the islands and indeed few places in the world attract so much rain. Consequently, the Hamakua Coast is a lush green paradise and an excellent place to propagate plants. Our explorations of the lush Hamakua Coast included two impressive botanical gardens. The Hawaiian Tropical Botanical Garden was an absolutely lovely spot that was full of thick and beautiful botanical specimens on lands lying adjacent to the ocean. Alas, though beautiful, most of the botanical specimens were non-native species like so much of present-day Hawaii. Nevertheless, it was an impressive collection of over 2000 distinct species that are classified into 125 botanical families and 750 separate genera. It was a joy to wander through the lush pathways of this garden down to the ocean.

A pathway at the Hawaiian Tropical Gardens leads past beautiful botanical displays to the sea

Visiting the nearby World Botanical Adventures Garden allowed us to stroll through its arboretum where my wife Vi saw many of her favourite trees from her childhood in Southeast Asia. Later, we walked through the Rainforest Trail past babbling brooks and intimate waterfalls. We also drove to an otherwise deserted overlook to view Kamaee Falls, a pristine thirty-metre waterfall that is replete with Hawaii's famous climbing fishes. It is only in this stream that one can find all seven of the freshwater fish that are native to Hawaii.[5] At least three of the goby species are capable of climbing up the waterfall rock faces to reach the stream above. Although waterfalls on the island carry seasonal run-off, which varies their appearance from season to season, the Kamaee Waterfall is different. The Kamaee Stream flows from a lava tube in which groundwater has been collected and thus the stream flow remains relatively constant throughout the year. The fact that the water has been sifted through the ground before entering the lava tube results in an untainted stream, remaining relatively clean even if storms have contributed a temporary boost in the strength of the Kamaee's flow.

The Kamee Waterfall at the World Botanical Gardens

A forest trail way at the World Botanical Gardens

Back on the road, we ventured north through the scenic countryside to the Waipi'o Valley, which abuts the Kohala Forest Reserve. The view from the Waipi'o lookout was awesome. Picture a verdant valley, and beyond the valley, high cliffs washed at their feet by the blue waves of the Pacific Ocean. That view certainly captured the magic and the beauty of the South Seas in my mind!

The Waipi'o Valley meets the ocean

We lunched in Waimea, where we stopped at a noodle shop and ordered a mix of noodles and vegetables in a salty broth that was a bit too salty for our taste. Afterwards, we drove out to the historic Parker Ranch. The Parker Ranch is one of the oldest in the islands, having been founded in 1847 by John Palmer Parker, and it was marking its 175th anniversary in 2022. The Parker Ranch memorializes the tradition of the *paniolos* – the famed Hawaiian cowboys, and the *pa'u* riders – early Hawaiian women who rode astride horses in long skirts when western women still insisted on riding side-saddle. I had a particular interest in visiting this ranch because it was the place where Princess Kaiulani spent her last happy days while celebrating the nuptials of her friend, Eva Parker before she contracted her fatal malady. But alas, I was to be disappointed because the Covid pandemic had closed the ranch temporarily. I was only able to inspect and photograph the entrance to the ranch and view some of the nearby mountains beyond the ranch where the Princess must have ridden on her last ill-fated excursion, caught and chilled by the cold Waimea rains and wind that the locals called the Spear of Waimea. The experience caused a breakdown in her delicate health and led subsequently to her death.

After briefly stopping at the ranch we turned back towards Hilo. On the way back we stopped at the small Laupohoehoe Train Museum. The train museum gave us an overview of the Hawaiian Consolidated Railway, a railway line once closely associated with the sugar cane industry that dominated the Hamakua coast in earlier days. The railway provided both freight and passenger train services along the coast but it was destroyed by the devastating 1946 tsunami, which also claimed the lives of 23 children and four of their teachers at a local school who were unable to escape the path of the wave.

Another day saw us returning to the island's west coast to visit some of the ancient and historic Hawaiian sites that the US government has chosen to preserve through its National Park Service. We visited *Pu'uhonua O Honaunau* National Park, which is also called the City of Refuge, a sacred place imbued with the *mana*, or spirit, of several great Hawaii *ali'i* who were interred there making this place a sanctuary where the ancient Hawaiians could escape retribution for breaking *kapu* or taboo. Behind a stone wall once guarded by Hawaiian soldiers in ancient days, the site remains a serene place that feels restorative to one's spirit. In old Hawaii, those who broke *kapu* often paid with their lives but if they could surmount the ocean breakers and scale the cliffs on the seaward side, or pass the guarded wall on the landward side, they would find safety within the refuge and the *kapuna*, or Hawaiian priests, would absolve them of their transgressions and allow them to return to their former lives free from the threat of religious retribution. The park today is a good place to see many marvellous and colourful fish in the pools along the shoreline, and there is a beautiful adjacent beach with picnic tables where we lunched afterwards.

There is a certain serenity and spiritual power at Pu'uhonua O Honaunau, the City of Refuge

Driving north along the coastal road we stopped to visit Kealakekua Bay State Park, also called by the locals K-Bay. Near here Princess Likelike, the sister of King Kalakaua, and the mother of Princess Kaiulani, once had a summer house and this is where she took Princess Kaiulani for a change of scenery away from bustling Honolulu in the days of the kingdom. I surmised that the pond where the young Kaiulani was said to have been allowed to swim was probably the pond that was constructed adjacent to the ruins of the ancient temple that stood on the beach. The young princess, who was the hope of the kingdom, and in the direct line for the Hawaiian throne, was not allowed to swim in Kealakekua Bay itself because of the presence and danger of patrolling sharks.

Kealakekua Bay today is a favourite spot for snorkeling and scuba diving because of the rich marine life that lives or traverses the bay. In the far distance across the bay, we could see the white obelisk erected by the Kingdom of Hawaii to mark the place where Captain Cook met his death. It was on the shore of this bay that the venerated naval explorer met his end at the hands of Hawaiian warriors in a dispute over a stolen boat. The monument itself lies on land that the Kingdom of Hawaii donated in perpetuity to Great Britain, and which in the past has been visited by ships of Commonwealth navies on friendly visits to the Hawaiian Islands.

Still travelling north we stopped next at *Kalaku-Honakahau* US National Park, which is the site of an ancient Polynesian coastal settlement possessing fishponds, temple ruins, and historic trails that are set amidst some ancient hardened lava flows. Some of the trails linked *makai* (the seashore), which was important for fishing, with *mauka* (the uplands), which was the setting for the planting and harvesting of crops. Of great importance in the early days was the Ala Kahakai trail that ran along the west coast of the island. Today, that historic trail is being restored and part of it traverses the seashore of *Kalaku-Honakahau*. As we walked along the shoreline that was part of the Ala Kahakai trail we could see several Green Turtles emerging in the surf to rest on the shore. The Green turtles are a protected species and signs warned us not to approach them within six metres. The ancient fishponds and the natural anchialine pools allowed for the mixing of freshwater and saltwater as part of the aquiculture practices of the ancient Hawaiians.

Ancient Hawaiian fishponds are preserved at Kalaku-Honakahau US National Park

Families relax on the beach at Kalaku-Honakahau US National Park

Further north but still on the west side of the island is *Pu'ukohola*, another US National Historic Site, this one honouring the deeds of the founder of the Kingdom of Hawaii, King Kamehameha the Great, who consolidated his rule over this island by erecting a great *heiau*, dedicated to the Hawaiian war god *Kuka'ilmoku*, or *Ku*. It was on the altar of this temple site that King Kamehameha I sacrificed his cousin Keoua, his main political rival for the lordship of the island shortly after the *heiau* was completed in 1791. *Pu'ukohola* apparently means the hill of a whale and indeed the massive site is very impressive. It was constructed from lava rock transported by hand by a chain of men from the Pololu Valley, some 32 kilometres away from the construction site. Here we learned that Kamehameha's rise to power had been foretold in three prophecies. First, there was passing by Earth of Halley's Comet in 1758, which was around the time when Kamehameha was thought to have been born. Secondly, the future king demonstrated his physical power by successfully throwing over the famous Naha Stone; a giant rock weighing 2500 kilograms, and finally, it was prophesied that if Kamehameha built *Pu'ukohola* he would rule the island of Hawaii.

The massive temple complex built by King Kamehameha I at Pu'ukohola

Kamehameha's conquest of Hawaii led to his subsequently successful military and naval campaigns to conquer Maui, Lanai, and Molokai, and then finally Oahu, which allowed him to formally establish the Kingdom of Hawaii in 1795. The Kingdom of Kauai and Niihau remained separate from the Kingdom of Hawaii however, but it was put under a constant threat of invasion by Kamehameha until the smaller kingdom willingly submitted to the sovereignty of its larger neighbour in 1810, in order to avoid an invasion. By 1810 King Kamehameha had created the consolidated Kingdom of Hawaii, which is the forerunner of today's US State of Hawaii.

We followed a footpath that took us down to the water's edge where the royal compound of Pekonu was formerly situated, and where an ancient *heiau* once existed that was dedicated to the Polynesian shark god. For centuries people living near here noted that sharks entered the murky, reddish, waters of the shallow bay in the daytime, which was unusual behaviour for sharks. Apparently, the sharks are still around because various signs explicitly warned us not to swim or wade in the water.

After a picnic lunch at the neighbouring Spence's Bay State Park, we visited the ancient petroglyph and the well-preserved Kalahuipua'a Fishponds that are still in use at the exclusive Mauna Lani Resort, where we also saw a variety of tropical fish, including a live ocean eel in the shallow waters along the coast. On the path to the beach resort, where we stopped for a drink before heading back to Hilo, I examined the historic Eva Parker Cabin, which was built as a retreat for that long-time friend of Princess Kaiulani by Eva's husband long before there was any thought of a resort in this area. The cabin has been preserved by the developers of the resort, and it is still used today for Hawaiian storytelling offered by the resort to its guests.

Back in Hilo, we drove to the Hilo Public Library to view the famous Naha Stone, a rather incongruous setting for this legendary artifact that we had learned about while visiting *Pu'ukohola*. According to Hawaiian legend, Kamehameha was able to lift or overturn the Naha Stone despite the fact that it weighs at least 2500 kilograms and possibly as much as 3500 kilograms (2.5 tons – 3.5 tons), thus proving his entitlement to claim the rule of Hawaii Island. King Kamehameha was reputed to be over 7 feet tall, and he must have been an extraordinarily powerful man if he was indeed capable of lifting that massive rock.

The famous Naha Stone that was lifted by Kamehameha at Hilo

One does not come to Hawaii Island without visiting, at least briefly, Volcano National Park, which harbours the famous Kilauea Volcano inside Halema'uma'u crater. The active volcanos of Hawaii Island include both the Kilauea Volcano and Mauna Loa but Kilauea is the main draw for tourism and the centrepiece of the national park. Kilauea is the most active volcano in the world. We visited it during the day and saw lots of steam rising from the caldera but there were no real lava flows visible during our visit. Later, we followed the Chain of Craters road through the park, visiting many of the important sites associated with past volcanic eruptions. The Thurston Lava Tube was probably the most interesting of the historic sites. That dark cave, now illuminated by dull lights, was formed when molten lava from the volcano flowed like a river through the rock creating the cave. When the lava flows stopped the cave remained. Discovered in 1913, the cave was named for the man who engineered the overthrow of the Hawaiian monarchy in 1893, Lorrin Thurston. Thurston was a lifelong promotor of the Kilauea Volcano as a tourist attraction and he was instrumental in obtaining national park status for the site in 1916. We continued following the Chain of Craters Road down to the ocean. Along the way, we were forced to stop when a pair of the rare Nene Goose crossed the road in front of us. This species is only found in the Hawaiian Islands and is considered to be endangered so substantial efforts are being made by the state government to protect it. Near the ocean, we were rewarded with fantastic views of the recent (2018) lava flows that had actually blocked the road. While I photographed the lava blockages Vi was also able to witness the magnificent sight of a humpback whale breaching from the ocean.

The Thurston Lava Tube in Volcano National Park

All of the Hawaiian Islands are the product of volcanic eruptions, with Hawaii Island currently being the youngest of the islands and the most volcanically active. But there is another island slowly rising from the floor of the Pacific Ocean. The Lo'ihi Seamount, also known now as Kama'ehuakanaloa is an undersea volcano lying some 35 kilometres southeast of Hawaii Island. Its last (underwater) eruption occurred in 1996. Although it is currently 975 metres below the surface of the ocean it is slowly rising and perhaps in another millennium, it will emerge from the sea into the bright sun that pervades all of the Hawaiian Islands.

South and east of Hilo an eclectic mix of people live in the area of the island known as Puna, a beautiful part of Hawaii Island but also a precarious area in which to live due to the possibility of eruptions from the nearby Kilauea volcano. In recent times no part of the island has been as affected by volcanic activity as Puna. In our exploration of Puna, our first stop was at the popular Maku'u Farmers Market, which is only held on Sundays. There, we found lots of reasonably priced fruits and vegetables, and at prices were considerably better than at the Farmer's Market that we often visited in Hilo. At the Maku'u market, we also found vendors hawking crafts and various bric-a-brac, such as old film cameras. There was also an assortment of plants for sale, and Vi saw a number of unusual orchids that she had never before seen offered for sale back in Canada.

After visiting the market we drove down Highway 132 to the ocean, stopping along the way at the Star of the Sea Church, a small Roman Catholic mission church that has been preserved by the local people. Inside, it is a painted church with a *trompe d'oeil* interior that was the artistic creation of the same Belgian priest who built the small wooden church a century ago. Much of the artwork in the painted church commemorates Father (later Saint) Damien, another Belgian priest who devoted his life to the lepers residing at the leper colony on the island of Molokai. It turns out that Father Damien was the original Catholic priest in the Kalapana area of Puna, which is close to where the church is now located, and Father Damien started the first Catholic church in the area, which he built with grass and leaves, later replaced by the current wooden structure.

The trompe d'oeil interior of the Star of the Sea Church in Puna

Further south lava flows had reached the ocean and we saw small houses built right on the lava terrain, which seemed to us to amount to a very hardscrabble type of existence. This whole area of the island is a preserve of various back-to-nature and off-the-grid people as well as local Hawaiian sovereigntists.

After retracing our steps we then took Highway 137 along the Pacific Ocean coastline, enjoying a lush and beautiful drive with gorgeous views of the Pacific Ocean. The lush vegetation was intermixed with pockets of the 'moonscape' terrain resulting from the 2018 lava flows. At Isaac

Hale Beach Park we stopped to take some photos of the recent hardened lava flows. At this beach, there was some notoriety due to alleged sightings of the famous night marchers, the ghosts of long-dead Polynesian warriors who are still searching for the location of their mortal remains. Some residents of Hawaii have reported seeing a procession of lights passing along ancient trails at night and these are said to represent the spirits of dead warriors. Of course, we visited in broad daylight so we did not run into the marchers, which was a good thing because it is said that if you do run into them at night you must avert your eyes and maintain a downcast demeanour. Otherwise, you risk death from the night marchers or so Hawaiian legend claims.

The effects of the recent lava flows in 2018 were very apparent to us and near the beach park the road ends suddenly in a wall of lava that entirely destroyed the road as the flow of lava marched into the Pacific Ocean. We partially retraced our path and then followed a back road to its junction with Highway 130 before heading back to Hilo. On our return, we also passed through the funky main street of the town of Pahoa, a significant but small municipality in Puna.

A wall of lava blocked a highway in Puna on the lava stream's 2018 march to the sea

As our departure from the island approached we drove once more across the island back to Kailua-Kona for a Covid-19 molecular test, a legal requirement for our re-entry into Canada during the Covid pandemic. We had been careful to avoid both crowds and indoor spaces during our visit and so, unsurprisingly, we both tested negative for the disease.

We had a picnic lunch on the West Coast at Kahlúa Beach Park where there is an ancient breakwall made of rocks that legend holds was constructed by the *Menehune* people otherwise known as the 'little people' of Hawaii. The *Menehune* were said to be expert craftsmen and several ancient structures throughout the islands are attributed to their workmanship. Supposedly, the Pao Ka Menehune's breakwater at Kahlúa Beach Park was constructed in a single night!

According to Hawaiian mythology, the leprechaun-like *Menehune* lived in remote areas of the islands away from other people and were only visible to Hawaiians who had a special connection to them. Some Hawaiians believe that the *Menehune* still exist in very remote areas of the islands. Although legends of small people can be safely relegated to mythology for the most part it is interesting to note that on the South Pacific island of Flores, in Indonesia, anthropologists have discovered a race of very small humans, *Homo floresiensis*, who received the nickname 'hobbit' because their physical stature did not exceed 1.09 metres (3 ft. 7 in.). Although no remains of such miniature humans have been recovered in the Hawaiian Islands the discovery in Flores seems to suggest at least the possibility of a race of small people on formerly remote islands.

Before departing Hilo for the last time I toured Wailoa State Park in central Hilo. Wailoa is a very picturesque park with ponds, a small river, and unique bridges with an undulating design. The state park is also the site of the state and county offices for the island and also contains war memorials, and a tsunami memorial to the Shinochi community, which suffered significant losses in the 1960's tsunami at Hilo. Perhaps its most significant monument is a large statue of King Kamehameha the Great, which has an unusual history. Apparently, the statue was commissioned for the town of Princeville on Kauai Island but it was donated instead to Hawaii Island because the locals on Kauai thought that a statue of King Kamehameha the Great would be inappropriate because Kamehameha never conquered their island! Instead, the Kingdom of Kauai had been peacefully ceded to King Kamehameha out of fear that he would invade it! The difference between conquest by brute force and peaceful cession under intimidation somehow escaped me but at least a fitting setting for the statue of this, the most historically significant indigenous Hawaiian, was found in Hilo.

On our final day on Hawaii Island, we left Hilo for the west coast again and took the coastal highway north along the Pacific coastline. Along the way, the highway the land rises considerably above the ocean, and that gave us tremendous scenic vistas. We stopped at Lapakahi State Park and saw the ruins of an ancient Hawaiian fishing village. Driving through Hawi to Kapa'ua we stopped to view the famous original statue of King Kamehameha the Great that was cast in Europe to commemorate the centenary of the arrival of Captain Cook on the islands. The statue had been ordered by King Kalakaua and his Premier, Walter Gibson, in the days of the Kingdom of Hawaii. This statue has an interesting history. After casting in Europe it was lost at sea during transport to Hawaii near the Falkland Islands in the South Atlantic. The government of the kingdom immediately commissioned a replacement for it and it was that replacement that was eventually erected in Honolulu in front of the Ali'iolani Hale, the Hawaiian Legislative Building (which is now the State Judiciary Building).

However, the original statue was later recovered from the Atlantic by salvagers and it was then obtained by Hawaii where it was decided that the recovered statue should be installed on Hawaii Island, somewhere in North Kohala near where the great king had been born and had spent his youth. The town of Kapa'ua was chosen to receive the statue and it was dedicated there in a formal ceremony presided over by King Kalakaua and unveiled by Princess Victoria Kinoiki Kekaulike, then the Governor of Hawaii Island, who had been born in Hilo. Today, much of the outdoor museum on the grounds reflects (deservedly) the sacrifices made by local men as part of the US armed forces in the twentieth century but no mention is made of the statue's original purpose or its unusual history. There is also a third, more modern, version of the King Kamehameha statue that was cast in the 1960s for inclusion in the US Capitol's Statutory Hall in Washington D.C. but that statue has now been relegated to the US Capitol Visitor Centre where the absolute monarch of the Hawaiian Islands rather incongruously greets Americans seeking to learn something about their democracy.

The statue of King Kamehameha I at Kapa'ua

From Kapa'ua we travelled down the scenic Kohala Mountain Road one of the most picturesque drives on Hawaii Island. On our way south in the early evening we stopped for a picnic at Anaeha'amolu Beach Park, a lovely beach set amidst a ritzy resort but nevertheless open to the public, as are all beaches in Hawaii. Then we drove down to Kailua-Kona to return our rental vehicle and to check in at the airport for our return flight back to Canada, which took the better part of February 2nd and required all of the now customary Covid-19 travel precautions, including masking in flight and within the airports. Upon arrival in Vancouver, I was randomly selected for a government-mandated Covid-19 PCR test which took a couple of days to process but did not interrupt our connecting flights back to Ottawa. (I subsequently learned that my test results were negative.) We arrived back home in the evening on a dark Ottawa winter night still wishing that we were back under a warm Hawaiian sky.

NOTES

[1] Melisse Malone, *Lost Bones of Kamehameha, Tyrone Young's 1983 Discovery*, Ke Ola Magazine, Keaau, HI, January-February 2022, page 16.

[2] Carole J Gariepy, *Meet Bernie Waltjen*, Ke Ola Magazine, Keaau, HI., November-December, 2021, p. 16.

[3] *Astronomy industry spent 110M in Hawaii in 2019*, Hawaii Tribune-Herald, Sunday, January 30, 2022.

[4] *Maunakea Plan Approved*, Hawaii Tribune-Herald, January 21, 2022, at pg. A-6.

[5] flagtail *Kuhlia xenura*, mullet *Mugil cephalus*, gobies *Awaous guamensis*, *Lentipes concolor*, *Sicyopterus stimpsoni*, *Stenogobius hawaiiensis*, and *Eleotris sandwicensis*.

14

A FINAL REFLECTION

At the outset of this travel memoir I described how a youthful mind had been captured by the stories of the South Seas, with its idyllic islands, broad sandy beaches, and the customs and ceremonies of an exotic people; an evanescent vision of the romance of the South Seas, once captured in the written works of explorers and novelists.

Of course, the Pacific of explorers and novelists could not exist unchanged by the late twentieth century but I still hoped to find a measure of authenticity when I was finally able to visit the islands of my youthful dreams. And I did find something of what those early observers wrote about in the quiet of a village on Yap, in the swirling colours of a Yap stick dance, in the beauty of a sunset on the beaches of Palau, the bejewelled emerald green of Palau's Rock Islands and the mysteries of Pohnpei's Nan Madol. But change was coming swiftly to these islands and on Pohnpei and Palau, I arrived at the moment of transformation, as two new island nations began to emerge from colonial obscurity to take their singular places in the world. At Pohnpei, I saw people using outrigger canoes that would not have been out of place a century or more before in the mysterious and ancient city of Nan Madol, while at nearby Palikir the finishing touches were being made to a new modern capital for this young nation. In both the FSM and in Palau the crest of a modern wave was washing over the islands in 1989.

Guam and Saipan were different. The indigenous cultures of the Chamorro people had been largely changed by 400 years of colonial rule by Spain, Germany, Japan, and the United States. By the late Eighties they were both thoroughly Americanized, Guam especially, and both were now part of the United States; Guam as a territory, and Saipan as a commonwealth, a sort of fancy name for an overseas possession. Like Hawaii, Guam and Saipan exhibited a type of regional American culture replete with fast food restaurants and fancy hotels.

On Saipan, there were bitter memories of the war amidst rusting tanks marooned offshore and the ruins of military fortifications. An irrepressible sadness descended over the ghostly settings of Banzai Cliff, and Suicide Cliff, where only the wind, and the tinkling of Buddhist memorials, disturb the silence where so many innocents perished.

And then there was Hawaii. The islands that the great American novelist Mark Twain had called "the loveliest fleet of islands that lies anchored in any ocean." More than any other islands in the North Pacific they have been changed beyond measure by the collision of a Pacific island culture and modernity. The nineteenth-century tragedy of the Kingdom of Hawaii can be summed up in a few words: disease, cultural domination, and economic capture. Today there are only 5000 pure-blood Polynesian Hawaiians left worldwide, according to the US Census Bureau. In recent decades a larger group of part-Hawaiians (although today's Hawaiians do not accept any distinction between those with a pure ancestry and those having a mixed ancestry) have joined together with the remaining Hawaiians to mount a cultural renaissance, a determined effort to recover the lost Polynesian heritage of the Hawaiian Islands. That renaissance is part of a worldwide movement by post-colonial indigenous peoples everywhere to recover something of what was taken from them through colonialism. But make no mistake, the Hawaii of today exhibits a regional American culture, and the islands have been thoroughly Americanized down to their roots. There is no authentic Polynesian population or Polynesian culture, still existing in the islands even though there has been a revival of some Polynesian customs, and some Hawaiian words are incorporated into everyday life in Hawaii. Most of the inhabitants of the islands today, or their ancestors, have come from the United States, southern Europe, or East Asia, and now there is a marked increase in migration from the American mainland. But much of the beauty of the Hawaiian Islands remains, and the memory of its vibrant Polynesian culture and of its unique political history is being preserved. Efforts are also being made to prevent the indigenous Hawaiian language from becoming extinct. Perhaps an element of 'Hawaiianess' will remain in the future, as a sort of Polynesian icing on an American cake.

My journeys brought me a better understanding of the vast Pacific world. Although much of the South Seas romance of the islands found in western literature has faded with the onrush of modernity these Pacific islands touched by America remain for me special places across the vast sweep of geography and time. In remote islands and in modern places I found Pacific peoples working to recover and memorialize their past, reconcile their traditions with modernity, and protect the beauty and wonderment of their idyllic paradises as a gift to a more prosaic world.

BOOKS BY PETER W NOONAN

Law and Legal History

The Crown and Constitutional Law in Canada (1998)

Sun Kinks and Saw Byes, Practising Transportation Law in the West in the Eighties (2012)*

The Crown and Constitutional Law in Canada, Second Edition (2017)

Canadian Boundaries, The Foreign Treaties And Other Instruments That Defined Our Realm (2018)

General History

Peace on the Lakes – Canada and the Rush-Bagot Agreement (2016)

Kaiulani Of Hawaii And The Fall Of Her Kingdom (2021)

Distant Allies, Canada and the Anglo-Japanese Alliance, 1900-1923 (2021)

Family History

The Shamrock And The Fleur De Lis: A Canadian Family Odyssey (2022)*

Travel

Journeys in the American Pacific: A Travel Memoir (2023)

* Privately published

www.ingramcontent.com/pod-product-compliance
Lightning Source LLC
Chambersburg PA
CBHW042147200426
43209CB00065B/1770